CLIMATE
IN
CRISIS

CLIMATE
IN
CRISIS

WHO'S CAUSING IT, WHO'S FIGHTING IT, AND HOW WE CAN REVERSE IT BEFORE IT'S TOO LATE

ROBERT F. KENNEDY, JR. AND DICK RUSSELL

FOREWORD BY DAVID TALBOT

Hot Books

Hot Books may be purchased in bulk at special discounts for sales promotion, corporate gifts, fund-raising, or educational purposes. Special editions can also be created to specifications. For details, contact the Special Sales Department, Skyhorse Publishing, 307 West 36th Street, 11th Floor, New York, NY 10018 or info@skyhorsepublishing.com.

Hot Books® and Skyhorse Publishing® are registered trademarks of Skyhorse Publishing, Inc.®, a Delaware corporation.

Visit our website at www.hotbookspress.com

10 9 8 7 6 5 4 3 2

Library of Congress Cataloging-in-Publication Data is available on file.

Cover design by Brian Peterson

Print ISBN: 978-1-5107-6056-1
Ebook ISBN: 978-1-5107-2176-0

Printed in the United States of America

Contents

This book is dedicated to the children
who will inherit the earth.

Introduction 2020:
The Deep State

By Robert F. Kennedy, Jr.

To greed, all nature is insufficient. —Seneca

Long before it turned its attention to systematically destroying the planet, the carbon industry set its sights on destroying American democracy and bulldozing our values.

The term "Deep State" is one of those toxic phrases that highlights and exacerbates the widening chasm between Democrats and Republicans. Ironically, polarization is a key strategic objective for the sinister cabal that the phrase describes.

Right-wing populists use the term "Deep State" to characterize the supposed authors of the cavalcade of social and economic demotions that have fatally wounded America's dwindling middle class. The obliteration has been so systematic and complete that it seems obvious to them that it is planned. In their view, a group of secretive aristocrats, led by George Soros and the late David Rockefeller, manipulate shadowy institutions like the Federal Reserve and the Council on Foreign Relations in order to shift wealth and power to billionaire elites, with the ultimate aim of achieving "World Government." To weaken the American character, their allies among the "Hollywood Elites" have deliberately cheapened our culture by infiltrating TV and film with sex and violence. These elites are purposefully undermining American democracy, subverting constitutional rights, and waging economic and cultural war on Main Street America, our democracy, and national sovereignty. Since all those cohorts have devoted energies to averting climate change, the global warming debate has become a prominent feature of these cosmologies; decarbonization is yet another attack on the US economy, and a ploy for One-World Government. Like every conspiracy theory, this one has nuggets of truth.

Democrats, meanwhile, dismiss talk of a "Deep State" as the ravings of deluded right-wing conspiracy theorists. They point out that Hollywood czars like Tom Hanks and Barbra Streisand are a great distance from real power, while arguing that Wall Street itself pushed sex and violence onto our TV screens; pornography, after all, sells sex.

David Rockefeller died in 2017, and the remaining Rockefeller family members have greatly diversified interests and little demonstrated *appetite* for One-World Government. They point out that George Soros is ninety years old, and manages vast investments in oil and gas with annual profits that dwarf his relatively tiny contributions to climate change activism. His Council on Foreign Relations (CFR) is an anemic think tank; its members mainly jockeying to burnish resumés and rub shoulders with a doddering Henry Kissinger. The CFR takes no position on foreign policy issues and mostly publishes informational white papers that gather dust.

But the Deep State does exist. It has already obliterated the middle class and has democracy on the ropes. The real power behind the curtain is a conglomeration of corporations, coal, oil, chemical, steel, and pharmaceutical--recently joined by telecom, Big Tech/Big Data—all bound, in a web of corruption, to our global military-intelligence apparatus. It is this collaboration of modern Robber Barons that is making war on democracy, civil rights, and the lower classes, while driving our country down the road to plutocracy and environmental apocalypse. This conglomeration has declared war on America's democracy and freedoms. Anyone who doubts that the Deep State exists should read the myriad histories of the CIA, including Tim Weiner's *Legacy of Ashes*, David Talbot's *The Devil's Chessboard*, Jefferson Morley's *The Ghost*, and James Douglass's *JFK and the Unspeakable*.

John D. Rockefeller laid the groundwork for the conglomeration with his ruthless drive for monopoly control of the world oil supply. (His company Standard Oil—now ExxonMobil—controlled 90 percent of the US oil supply.) The pharmaceutical cartel is the offspring of Rockefeller's US petroleum and coal tar industries and Third Reich chemists, who were deeply incriminated in the Holocaust and the Nazi war effort. As William Shirer documents in his book, *The Rise and Fall of the Third Reich*, it was Germany's oil and coal industry and the pharmaceutical giant I.G. Farben that principally financed Hitler's rise to power. Rockefeller gained controlling shares in IG Farben (now Bayer, the German chemical and pharmaceutical conglomerate). His philanthropy focused on his philosophy of promoting petroleum-based pharmaceuticals and marginalizing the previously popular alternative medicines: osteopathy, homeopathy, natural remedies, and plant medicines. For decades, the Rockefeller family owned some 80 percent of the US pharmaceutical industry. Today, the Rockefeller Empire—in tandem with JP Morgan Chase—continues to own half the pharmaceutical industry in the US. John D. Rockefeller's grandson, David, with his ties to the oil and pharmaceutical industries and international banking, and his friendship with Allen Dulles, was certainly among the Deep State's Grand Panjandrums.

David Rockefeller used his close relationship with the CIA, initially through Dulles, and his position as a director and then chairman of the CFR, to wage war on nationalist movements and representative democracies around the globe whenever they threatened the profits of his oil, mineral, chemical, pharmaceutical, or banking interests.

All of these histories reveal Big Carbon as the conspiracy's center of gravity. The CIA and the military nurtured a long cozy relationship with King Coal and Big Oil. Since the abolition of slavery, the Navy's replacement of sailboat fleets with fossil fuel–driven transportation, and the introduction of petroleum-based pharmaceuticals, American wars have been, to a greater or lesser extent, strategic struggles over control of coal ports, shipping routes, and oil fields. America's first great foreign excursion was Cuba's War of Liberation. In 1898, America's Yellow Press appealed to the nation's idealism to drum up popular support for US intervention, purportedly to support Cuban revolutionaries in their struggle for independence from Spain. To create a pretext for intervention, Deep State militarists staged the "false flag" scuttling of the battleship, the *USS Maine* in Havana harbor, and blamed the sinking on Spain. Their true objective became clear soon after Cuba's independence fighters achieved victory over Spain. Deep State militarists robbed the new nation of its most important port, expropriating Guantanamo Bay as a Naval coal terminal. A century later, Guantanamo, the symbol of America's abandonment of its revulsion for imperialism, became the situs for where America rebuffed its seminal revulsion against torture. Today, Guantanamo Bay stands as a kind of "anti-Statue of Liberty"—a hemispheric symbol of the surrender of fundamental American values to the Deep State military-industrial complex, with its devouring hunger for carbon and its ambition for authoritarian control. After Cuba, protecting American oil and coal resources and ports became a *raison d'être* for an endless parade of new American wars and interventions.

A nation's political structure tends to reflect its economic organization. When major industries are owned and controlled by a small group of wealthy individuals, the nation itself becomes economically stratified and tyrannical. Coal and oil are authoritarian industries. They are highly capitalized and rely on ruthless control of real estate and resources. While most nations declare that their oil resources are publicly owned, in practice poorer citizens rarely share in petroleum profits. Large multinationals, frequently allied with local oligarchs, invariably and systematically find ways to steal and monopolize these resources.

The term "oil curse" describes the nearly universal dynamic by which the governments of nations or states with rich oil reserves invariably devolve into highly militarized and despotic organs that are brutal and dictatorial. Oil-dependent economies generally foster giant wealth gaps between rich and poor and violent totalitarian governance. The strategic relevance of oil and steel make these industries natural allies of the military and intelligence apparatus.

In 1954, the CIA's director, Allen Dulles, overthrew the democratically elected government of Iran, after the country's beloved Prime Minister, Mohammad Mossadegh—the first democratically elected Head of State in Persia's 4,000-year history—committed the "crime" of threatening to nationalize oil fields controlled by BP and Texaco. (Texaco had formerly been Dulles's client at the white-shoe law firm, Sullivan & Cromwell.) Dulles installed the Shah to rule Iran and to protect the financial interests of his petroleum company client. Seventy years later, the entire world is still suffering from the blowback of that coup: the Iran Hostage Crisis of 1979 (precipitated when David Rockefeller and his crony Henry Kissinger, pressured President Jimmy Carter to host the deposed Shah of Iran in the United States, in order to protect Chase Manhattan's assets); the rise of militant Islam; the Afghanistan, Iraq, and Syrian war (which flooded Europe with displaced refugees, undermining Europe's unity and democracies); and our continued enmity with Iran, a nation that, in every other respect, should be our closest Middle East ally—are all blowback from that anti-democratic coup.

The founders recognized that America could not be both an imperial power abroad and a constitutional democracy at home. It is a seminal axiom of American foreign policy that our democratic government should not embroil America in foreign wars. President John Quincy Adams summarized the consensus of the founders, when he declared,

[America] goes not abroad, in search of monsters to destroy. She is the well-wisher to the freedom and independence of all. She is the champion and vindicator only of her own. She will commend the general cause by the countenance of her voice, and the benign sympathy of her example.

But the American carbon titans successfully overwhelmed these qualms and deployed the US military as their private army, expanding its reach to protect Big Carbon's global interests.

Allen Dulles helped his oil-fueled Deep State cabal to engineer America's departure from its traditional principle of non-intervention. My grandfather picked our family's sixty-plus-year fistfight with the CIA in 1954 when President Eisenhower appointed him to a commission, chaired by former president Herbert Hoover, investigating the CIA. The official tenet of American foreign policy was supporting the spread of democracy. Yet the Hoover Commission found that the CIA was working, in league with petroleum companies, in direct opposition to official US State Department policy, and regularly engaged in conduct antithetical to US values. Following President Adams, my grandfather believed that America could not be at once an imperial power and a constitutional democracy. He was angry and disgusted to learn that Dulles and his agency were overthrowing governments, disrupting elections, bribing politicians, and undermining democracy around the globe in service to Big Oil and mining companies, as well as industrial/chemical agriculture. The CIA orchestrated changing other countries' governments seventy-two times during the Cold War, affecting nearly one-third of the nations on Earth.

I was in Chile in 1973 during the coup d'état orchestrated by the CIA and David Rockefeller for the sole purpose of protecting US telecom (ITT), banking (Chase), oil (Texaco), chemical (DuPont and Dow), food (PepsiCo Inc), and mining interests (Anaconda)—all threatened with nationalization by President Salvador Allende. I was fired upon—and nearly killed—by an army patrol as I struggled on foot across the Andes to Argentina. I was very conscious of the key role played by David

Rockefeller and the US Telecom behemoth ITT Corporation (International Telephone and Telegraph).

Strategic lust for oil and the need to protect oil infrastructure motivated most of the CIA's interventions. CIA spooks and paramilitary units often worked hand in hand with mercenaries and private armies, which American oil companies paid, armed, and trained. Foreign warlords and politicians, fattening on oil company payrolls, routinely sold out the interests of their own nations (and murdered their own citizens) in furtherance of petroleum company ambitions. They financed and trained tens of thousands of oil industry personnel throughout the Middle East as paramilitaries, to fight off the Soviets and destroy oil infrastructure to prevent its capture in the event of Soviet invasion. According to biographer David Talbot, Dulles was incapable of distinguishing the US national interests from the interests of his former oil industry clients. In fact, US national interests took a distant back seat to shareholder profits.

My grandfather recommended the disbandment of the CIA's "Plans" division. He feared that the dangerous alliance of military and intelligence apparatus with oil industry fat cats would have disastrous consequences for America's democracy and our global reputation, potentially turning America into a National Security State. He recognized that the large oil corporations have no loyalty to America, much less to our core values. A statement by Exxon CEO Lee Raymond, during a 1998 ExxonMobil meeting, confirmed my grandfather's suspicions about the essential lack of patriotism among oil men: "I'm not a U.S. company," stated Raymond, "and I don't make decisions based on what is good for the U.S."[1]

It is no accident that the Deep State kingpin, Charles Koch (founder of Koch Industries, the largest privately owned oil company in the world), made his fortune by building refineries for the homicidal communist dictator Joseph Stalin. Koch and his sons deployed billions of ill-gotten dollars to create an infrastructure of Deep State thinktanks like the Heritage

---

1   Coll, Steve: *Private Empire: ExxonMobil and American Power*, 2012, Penguin Books.

Foundation, the Cato Institute and the Competitive Enterprise Institute, tasking them the job of creating the philosophical underpinnings for the domination of America's democracy by corporations and militarists.

On my seventh birthday, January 17, 1961, three days before my uncle John F. Kennedy's inauguration as America's 35th president, outgoing president Dwight D. Eisenhower delivered the greatest speech of his career, warning Americans against the Deep State, which he called the "military-industrial complex."

Eisenhower cautioned that the cartel could destroy our democracy.

This conjunction of an immense military establishment and a large arms industry is new in the American experience. The total influence—economic, political, even spiritual—is felt in every city, every Statehouse, every office of the Federal government... [W]e must not fail to comprehend its grave implications. Our toil, resources and livelihood are all involved; so is the very structure of our society.

Eisenhower went on:

In the councils of government, we must guard against the acquisition of unwarranted influence, whether sought or unsought, by the military-industrial complex. The potential for the disastrous rise of misplaced power exists and will persist.

He cautioned that Americans must learn to recognize, and guard against, all the milestones of tyranny.

We must never let the weight of this combination endanger our liberties or democratic processes. We should take nothing for granted. Only an alert and knowledgeable citizenry can compel the proper meshing of a huge industrial and military machinery of defense with our peaceful methods and goals, so that security and liberty may prosper together.

On the same day as Eisenhower's speech, just days before my uncle took the oath of office, Belgian intelligence spies—with the support of Allen Dulles's CIA—murdered the Congo's charismatic leader, Patrice Lumumba. As a US Senator, JFK had used his position as Chair of the U.S. Senate Sub-Committee on African Affairs to support Lumumba and other anti-colonial leaders. Dulles was disdainful of my uncle's sympathies for African liberation movements and, in particular, his admiration for Lumumba. Dulles wanted to commit this murder before JFK took office. The Congo ranked among the world's richest nations measured by its mineral wealth and natural resources, including its abundant oil fields. US and European mineral and petroleum companies were salivating at the prospect of exploiting tribal tensions to cut the newly liberated nation into bite-sized parcels that they could easily devour and rule. They knew Lumumba was the only Congolese leader with the charisma and popularity to unite all of Congo's rival warring tribes. Lumumba's murder shocked and saddened my uncle. He would not live to learn of the CIA's role in orchestrating that hit for Big Oil and the Deep State.

Beginning in 1958, Dulles had worked with Eisenhower's bellicose vice president, Richard Nixon, plotting to oust Cuba's newly ascended revolutionary leader, Fidel Castro, who earlier that year had deposed brutal dictator Fulgencio Batista and his mafia cronies. Nixon and Dulles persuaded Texaco to shut down its critical Cuban refinery, and United Fruit—another of Allen Dulles's former clients—to cease exports of Cuban sugar, so as to crush their economy and destroy Castro's revolutionary regime. Castro and his lieutenant, Che Guevara, were avowed Marxists, but their revolutionary colleagues embraced a wide range of competing ideologies that were mainly democratic and anti-Batista. The CIA's preemptive strikes, bent on starving the tiny nation, forced Cuba to turn to Russia for financial help. The Soviets agreed to rescue the besieged Cubans by trading Russian oil for Cuban sugar. The CIA retaliated with the aerial bombing of a Havana shopping center. That illegal CIA act of terror gave Castro the political strength to declare his new regime communist for the first time. He made the announcement at the funeral of the CIA's victims. In 1989, Castro told me, "The United States facilitated Cuba into embracing Marxism."

When JFK denied the CIA's request to transport the "Bay of Pigs" brigade in naval vessels, Dulles's former client, United Fruit Company, provided the CIA with a fleet of ships to support the invasion. When Castro's overwhelming forces predictably trapped the brigade on the beach at Playa Girón, JFK refused the CIA's request for air support by US forces. Dulles had assured Jack that US military intervention would, under no conditions, be necessary. JFK realized that Dulles and other CIA officials and military brass had lied to him. He told his closest advisors, "I want to splinter the CIA into a thousand pieces and scatter it to the winds."

JFK fired Allen Dulles after the Bay of Pigs, but Dulles continued to steer the CIA remotely and would return to government in 1963 to lead the Warren Commission investigation of JFK's death. He used that post to conceal the CIA's deep involvement in JFK's assassination.

My uncle and father spent their careers—and gave their lives—to the task of saving democracy from the Deep State cartel. They enraged Dallas oilmen by their efforts to revoke the Oil Depletion Allowance and other tax subsidies, which then provided corporate welfare worth $185 million, annually, in handouts to America's petroleum corporations. Texas oil producers paid zero income tax on 27.5 percent of their taxable income. A handful of powerful Dallas oilmen, most notably Clint Murchison, H. L. Hunt, Sid Richardson and D. H. Byrd, were making millions from the allowance. Byrd alone had a $30 million annual income (in the 1960s) and paid no federal taxes. These Welfare Cowboys all had close connections with the CIA.

Big Steel also had a natural alliance with the Pentagon. In 1962, JFK mediated a fierce wage dispute between the United Steel Workers' Union and the twelve largest steel companies. The president intended to hold inflation in check by fighting the wage/price spiral. His delicate negotiations led to an extraordinary agreement: workers acquiesced to a wage freeze and the steel companies agreed to freeze steel prices. Shortly after signing the agreement, US Steel CEO, Roger Blough, came to the Oval Office to announce that Big Steel was committing a double-cross. The top six steel companies were violating the agreement and unilaterally raising prices by $6/ton (3.5 percent). Jack told Blough, "You are making a big mistake." He then ordered the Pentagon to move ship building contracts from the noncompliant companies to smaller companies that had not raised prices. My father, Attorney General Robert Kennedy, sent FBI agents to raid the big six steel company offices and to cart out their filing cabinets on dollies. His deputies told the steel executives to expect prosecutions on tax evasion and anti-trust activities. It was the strongest pushback by a US president against corporate power since Andrew Jackson fought the banks. The *Wall Street Journal* denounced JFK for his naked power play against Big Business. Wall Street and the Deep State never forgave him. But that was only the beginning of Camelot's war on the Deep State.

JFK defended Rachel Carson, whose book *Silent Spring* resulted in the banning of DDT, against Monsanto and the chemical conglomerates. He sent Pentagon and CIA spooks into apoplexy when he refused to dispatch combat troops to Laos, Vietnam, or Cuba. When he proposed détente with Khrushchev and Castro, Nelson Rockefeller accused him of treason. Jack signed the nuclear test ban treaty that the oil industry violently opposed; petroleum companies were America's largest producers of uranium, and they feared that peace would undermine their business model.

My uncle Jack died two months after signing the Atmospheric Test Ban treaty, and 14 weeks after signing National Security Order 267, which ordered all US advisors out of Vietnam by December 1965. JFK's death saved hundreds of millions of dollars for Texas oilmen by permanently deferring his plan to repeal the Oil Depletion Allowance.

My father's first instinct was that the CIA had murdered his brother. Less than two years later, LBJ had ordered close to 250,000 US combat troops into Vietnam, converting that country's civil conflict into an American War, in which more than 58,000 Americans and millions of Vietnamese would die. In 1968, my father died during a presidential campaign waged against the war machine. He promised he would end the Vietnam War on the day he took his oath of office. He told his writer friend, Pete Hamill, that he meant to break up the CIA. Three weeks before his murder, he publicly acknowledged that he intended to reopen the investigation of his brother's assassination. His death was one of five great national tragedies—the assassinations of JFK and Martin Luther King Jr.; the Vietnam War; 9/11; and COVID-19—that allowed the men who want permanent war to transform America, which was once the world's exceptional democracy, into a National Security State.

JFK and RFK dedicated their careers to preserving American democracy and idealism. One of their defining struggles was against the Deep State, and in particular the clawing power of Big Carbon. My own career has continued this tradition.

In the 1970s, Exxon, formerly Standard Oil, employed the world's most brilliant carbon scientists to understand every stage of oil production. The company prided itself on knowing more about the fate of the carbon molecule than any individual, corporation, or government on Earth. Internal documents, created in the years following my father's death and recently made public, show that those scientists warned Exxon that business as usual would melt the polar ice caps, cause sea levels to rise, and trigger cataclysmic climate change. Exxon CEO Raymond described "so-called global climate change" as "the issue that perhaps poses the greatest long-term threat to our industry."[2] Rather than change its business model to save humanity, Exxon and its Carbon cronies invested half a billion dollars in a four-decade campaign of lies and deception to gull the public into

---

2   C-SPAN, Chairman's Address at the American Petroleum Institute Annual Luncheon, November 11, 1996.

believing the absurd: that climate change is a hoax. A century earlier, one of my father's favorite poets, Rudyard Kipling, described such deception as truth being "twisted by knaves to make a trap for fools." Big Oil's Deep State cronies—the intelligence apparatus, the military-industrial complex, and some mainstream media, who take cues from oil and automobile advertisers—abetted this venal campaign.

Two forces drive democracy: money and political intensity. The Deep State has money but its mission—to further enrich and empower the wealthiest 0.001 percent—is not a potent vessel for populism. Instead, it relies on so-called "wedge" or "culture war" issues to recruit foot soldiers. Steel, Oil, and Pharma deploy front groups and PR flaks, using all the alchemies of demagoguery: dog-whistle racism, bigotry against immigrants and people of color, and invocations to patriotism and Christianity, in order to engage those whites who still feel alienated by the Civil Rights laws of the 1960s. The Wall Street fat cats, the power brokers, and PR flaks talk openly about "gulling the rubes" in their rural redoubts, winning their loyalties with palaver of abortion and the Three Gs: God, guns, and gays. In the words of Christian Coalition President Ralph Reed, "They write the checks, and they get the joke."

These cohorts have used their money and political power to engineer for themselves obscene subsidies and tax breaks. (Exxon, the wealthiest company on earth during the 1990s, paid virtually no federal taxes.) According to the World Bank, Big Carbon receives annual subsidies of $5.4 trillion globally and $655 billion in the US—more than we spend on our military and ten times our education budget. Using political clout, the carbon incumbents have written the laws that regulate energy in America to reward the dirtiest, filthiest, most tyrannical and warmongering fuels from Hell, rather than the cheap, clean, green, democratic, and patriotic fuels from Heaven.

The carbon cronies have made climate change a defining feature of that conversation. Flipping the narrative, they portray climate change advocacy as a sinister effort to establish world government and rob America of its economic independence and sovereignty. Big Carbon's big lie is that any change in the *status quo* would raise gas prices and deplete middle-class jobs.

I've spent forty years fighting to stop the oil and coal titans from contaminating our water, our air, and our children with toxics, arsenic, benzene, PAHs, the mercury that has poisoned every freshwater fish in America, *and* the carbon that now poses an existential threat to our planet. I have litigated these issues across the Americas. As a partner of the clean tech investment firm, Vantage Point Venture Partners, and advisor to Stanwool Energy, I was involved in building transmission lines and generation infrastructure for clean, democratic energy from wind and solar, including the two largest solar plants in North America. I've been deeply involved in building and deploying renewable energy infrastructure that competes with oil and coal. Vantage Point was the earliest investor in Tesla and the force behind Ivanpah, the world's largest solar thermal plant. I know the science and the economics of carbon energy backwards and forwards.

Predatory industries always employ the same playbook. Big Tobacco swore smoking didn't cause cancer. Monsanto convinced us that DDT and glyphosate were harmless. Pharma lied to persuade doctors and the public that opioids were not addictive, that Vioxx did not cause heart attacks, that vaccines are undeniably safety tested, and that the autism epidemic is an illusion. They use the same phony (tobacco) scientists and mercenary biostitutes to gin up fraudulent studies that sow doubt, paralyze policy reform, and give political cover to their tame politicians. They all work together in lockstep, coordinated by their Capitol Hill trade associations, lobbying firms, captive agencies, and paid-off politicians to increase authoritarian control, to transform all of us into mindless consumers, to shift middle class wealth to billionaire plutocrats, and to liquidate our "*purple mountain majesties*" and our entire planet. They enrich themselves by impoverishing the rest of us. They capture regulators, seduce reporters, corrupt science, and pay off lawmakers to subvert democracy. They employ state-of-the-art propaganda, psychological warfare, and all the formulaic alchemies of demagoguery to divide us. For forty years, I've worked with the Left on conservation, climate, energy, and the environment. In recent years, in my battles against Pharma, I have also worked with many right-wing allies, including Trump supporters.

Big Oil, King Coal, and Big Pharma are all titans of the Deep State cartel of deception and authoritarian control. In its most audaciously ruthless betrayal, this cartel has engineered a suicide pact for humanity and our planet. They have brought us to the eve of Armageddon. Their business plan poses an existential threat to humanity. They have declared war on democracy and personal freedom. They are the Four Horsemen—the apocalyptic forces of ignorance and greed, pestilence and fear.

Today we are living in the beleaguered world that the Deep State's greed and negligence has created—the science-fiction nightmare that these criminals devised. The glaciers are melting on every continent, threatening the food and water supplies for billions of people. The ice caps are shrinking, their melt water swelling the oceans and flooding coastal cities. Fisheries are collapsing globally. Disease, drought, fire, famine, and flood are transforming the planet into scenes reminiscent of Biblical accounts of the Apocalypse.

This is not a hoax. You don't need a science degree to know the planet is warming any more than you need a degree to know that the nation's autism epidemic is real. In both cases, you need to be deliberately blind to ignore the evidence.

Within the short time span since the first edition of this book (then titled *Horsemen of the Apocalypse*) appeared in 2017, the pace of climate change has accelerated beyond what anyone had then foreseen. July 2019 became earth's hottest month since recordkeeping began in 1880.

I live on the West Coast now, where the fire seasons in California are two months longer than they historically have been. Twice in the past two years, my family has had to evacuate our home, in an area that was never part of the traditional fire zone. My family has a summer home on Cape Cod. Two 100-year storms struck our town over the past two years, destroying a pier that had withstood every storm for a century.

These are my personal harbingers of the predictable fallout of climate change: storms on steroids, droughts, famine, disappearance of the ice caps and the glaciers on every continent, spread of insect-borne tropical diseases, threatening civilization and humanity. Wildfires raged across Alaska, the Arctic, Greenland, and Siberia; Australian wildfires devastated that continent; California wildfires are 500 percent larger than they were in the 1970s; the burning of the Amazon is leading to a cascading collapse of natural systems across the planet. In the world's second-largest rainforest, the Congo, 50 percent more fires burned than in the Amazon.

The Canadian Arctic permafrost is thawing seventy years sooner than had been predicted, and a UN report warns that at least 30 percent of the Northern Hemisphere's permafrost will melt during our children's lifetime, creating feedback loops that will release billions of tons of methane, the worst of the greenhouse gases. During 2019, sea ice within 150 miles of Alaska's shoreline completely melted away for the first time in recorded history. The state's largest city, Anchorage, baked in temperatures of 90 degrees Fahrenheit. Salmon died of heat stress, and levels of shellfish poisoning soared. Then, the Greenland ice sheet lost a mind-bending 12.5 billion tons of water in a single day. That ice sheet contained enough frozen water to raise sea levels around the world by 20 feet. Hundred-year flooding events are now a regular occurrence in the US, especially in the Northeast and Southeast. Superstorms and rising sea levels will displace as many as 280 million people around the world. A US report warns that twenty-one beach towns—including Miami Beach, Galveston, Atlantic City, and Key West—will soon be underwater. Indonesia is already relocating the millions who live in its capital city of Jakarta. When I visit our Great Lakes Waterkeepers, I see steadily rising water levels that have put the Great Lakes communities in crisis. Both Detroit and Miami mayors have declared states of emergency as rising waters threaten their cities.

Even as floods are drowning our great municipalities, meteorologists are predicting megadroughts on a scale last witnessed in medieval times. During a 2019 heat wave that swept across Europe, 1,500 people died from heat stroke in France alone. Desert heat will parch the US Southwest within decades. The UN estimates that two billion people are already facing moderate to severe food insecurity, primarily due to the warming planet. Warmer oceans and a hotter, wetter atmosphere provide steroids to storms. I visited with Bahamas Waterkeepers before and after Hurricane Dorian, which, in 2019, wiped out thousands of housing compounds and businesses on Grand Bahamas. Many of these had taken Bahamian families three generations or more to create.

In August 2019, with global atmospheric CO_2 levels already at 415 ppm, Iceland citizens held a funeral for the once massive Okjokull glacier, and erected a plaque which reads:

A letter to the future. Ok[jokull] is the first Icelandic glacier to lose its status as a glacier. In the next 200 years all our glaciers are expected to follow the same path. This monument is to acknowledge that we know what is happening and what needs to be done. Only you know if we did it.

Nowadays, every week seems like a new chapter from the Doomsday Book of Revelation, but I will close with a warning penned by a desert prophet a thousand years earlier.

The earth lies defiled
under its inhabitants;
for they have polluted the land, transgressed the laws,
violated the statutes,
broken the everlasting covenant.
Therefore a curse devours the earth,
and its inhabitants suffer for their guilt;
therefore the inhabitants of the earth are scorched,
and few men are left....

—*Isaiah* 24:5-6

If we are to avoid the curse, we need to call out the authors of the sinful pollution. As the late activist musician Utah Phillips observed more than a decade ago: "The earth is not dying, it is being killed, and those who are killing it have names and addresses."

The Deep State profits from our division and employs so-called "bourbon strategy" techniques to keep Americans in internecine struggles: black against white, urban vs. rural, Christian against Muslim, right against left, blue collar vs. white collar, Republican against Democrat; to distract us all from waging class war against the Deep State Elites. The term should, instead, unite us.

It's time to find our common ground and fight the real Deep State, and not each other. We need to work together to escape the seduction of their lies and propaganda so we can unite against the real villains!

Prelude 2020: The Youth Uprising and What They're Up Against

By Dick Russell

On September 24, 2019, a few days after protests across the globe had mobilized four million people in 163 countries, sixteen-year-old Greta Thunberg stood before the United Nations Climate Action Summit in New York City. She'd begun sitting outside the Swedish Parliament a year earlier, holding up a sign that translated "School strike for climate." Now she'd sailed across the Atlantic in a 60-foot racing yacht equipped with solar panels and underwater turbines; a carbon-neutral crossing to demonstrate the importance of reducing emissions.

Thunberg stood before the UN delegates and told them: "You have stolen my dreams and my childhood with your empty words. And yet I'm one of the lucky ones. People are suffering. People are dying. Entire ecosystems are collapsing. We are in the beginning of a mass extinction, and all you can talk about is money and fairy tales of eternal economic growth. How dare you!"

It was a stunning moment, as Thunberg accused world leaders of ignoring thirty years of scientific warnings. "The popular idea of cutting our emissions in half in ten years only gives us a 50 percent chance of staying below 1.5 degrees [Celsius] and the risk of setting off irreversible chain reactions beyond human control. Fifty percent may be acceptable to you. But those numbers do not include tipping points, most feedback loops, additional warming hidden by toxic air pollution or the aspects of equity and climate justice. They also rely on my generation sucking hundreds of billions of tons of your CO_2 out of the air with technologies that barely exist."

Hers and future generations would be living with the dire consequences of world leaders' inaction. "You are failing us. But the young people are starting to understand your betrayal....We will not let you get away with this. Right here, right now is where we draw the line. The world is waking up. And change is coming, whether you like it or not."

Thousands Who Demand We Do It:

According to a survey taken by Amnesty International in December 2019, young people in 22 countries identified climate change as the most important issue facing the world. They are the future. They hail from all parts of the world, all backgrounds, all ethnicities. Many have already experienced firsthand the devastating effects of climate change. Some are mobilizing politically behind the Green New Deal introduced by Rep. Alexandria Ocasio-Cortez and Senator Ed Markey, a ten-year plan to achieve net-zero greenhouse gas emissions. Others are taking to the streets with acts of civil disobedience. Still, more are involved in lawsuits against federal and state governments, or pushing college administrators to divest their assets from fossil fuel companies.

Fourteen-year-old Alexandria Villaseñor from New York, one of the youngest organizers of the Global Climate Strike, is the founder of Earth Uprising. She'd visited families in 2018 during the Camp Fire in Northern California—deadliest in the state's history—where the smoke triggered her asthma. Researching the wildfires, she'd learned how climate change is helping ignite them. She and five other young activists have filed a legal action against the five countries which are considered the biggest contributors to climate change, on the premise that this violates the UN Convention on the Rights of the Child.

Vic Barrett, a twenty-one-year-old from White Plains, New York, is an Afro-Latino activist originally from the island of St. Vincent. After Hurricane Sandy left his family home and school without power for days in 2012, he became a Fellow with the Alliance for Climate Education and part of the landmark lawsuit filed in Oregon, *Juliana vs. United States*, suing the government—along with twenty-one other young plaintiffs—for its role in the climate crisis. (In mid-January 2020, the 9th US Circuit Court of Appeals dismissed the case, with two judges maintaining that despite "a compelling case that action is needed....the panel reluctantly concluded that the plaintiffs' case must be made to the political branches or the electorate at large.")

Xiye Bastida, seventeen, was forced to leave her hometown in San Pedro Tultepec, Mexico, after floods kept her from attending school. Relocating with her family to New York City, she learned about Hurricane Sandy and realized that climate change is a global problem. She became a Fridays for the Future NYC organizer, and youth ambassador for the Y on Earth nonprofit that provides communities with information on sustainability.

Benji Backer, a native of Wisconsin, is a twenty-one-year-old senior at the University of Washington in Seattle and president of the American Conservation Coalition, launched to emphasize market-based environmental reforms. Along with Thunberg and Barrett, he addressed a congressional committee in October 2019.

Jamie Margolin, a high school senior in Seattle who co-founded the Zero Hour organization in 2017, also spoke before Congress that year. Her catalyst was the abysmal government response to Hurricane Maria's devastation of Puerto Rico, and her own difficulties breathing due to wildfires in Canada. She's one of the plaintiffs suing the state of Washington for inaction, who maintain that a stable climate is a basic human right. In her fiery speech following the UN Climate Summit, Margolin accused the Congress of "promising me lies, lying to my face about a bright future ahead." She is committed to "transforming our culture into one enabling radical climate action."

Evan Weber, twenty-six, is a co-founder of the Sunrise Movement. Originally from Hawaii, where he watched rising sea levels wipe out several of his favorite childhood beaches, his organization launched the Sunrise Movement with some grant seed money from the Sierra Club and Bill McKibben's 350.org. In November 2018, two hundred Sunrise activists occupied House Speaker Nancy Pelosi's office, demanding action on climate; more than fifty were arrested. A month later, Sunrise returned to Congress with a thousand young people. The group now has nearly three hundred "hubs" around the country. Many have begun to not only work but live together in "movement houses."

Isra Hirsi, sixteen, is the daughter of Minnesota US Representative Ilhan Omar. She got involved when she saw how much climate change was affecting people of color. Hirsi is now the executive director and co-founder of the US Youth Climate Strike, organizing walkouts around the country and advocating for the Green New Deal.

Globally, four out of ten people are below the age of twenty-five. Over a thousand came together in September 2019 at the UN's Youth Climate Summit, representing 140 different countries. Their presence caused UN Secretary-General Antonio Guterres to speak of how the discouragement and "sense of apathy" he'd been feeling changed "largely due to the youth movement." Guterres told the gathering, "We are still losing the race but are finally starting to see governments reacting. It is not only a question of glaciers or ice caps or corals, but more and more about the suffering of people....I want my grand-daughters to have a liveable [sic] planet. Your generation must make us accountable, to make sure that we don't betray the future of humankind."

Bruno Rodriguez, a nineteen-year-old from Buenos Aires, Argentina, spoke of the "existential emergency" and the "new collective consciousness" forming around it. "There are no frontiers to fight for structural change," he said. "We will not wait passively. The time is now for us to be leaders."

Youth for Nature has collected eighty stories from 85 countries, about Nigeria teaching local farmers how to practice agroforestry techniques, Brazil holding reforestation workshops in Amazonia, and restoring coral reefs in the Caribbean. Other nature-based solutions include a "Great Green Wall of Africa," an $8 billion tree planting project in the Sahel. In Kenya, youths with drug problems are being employed as tour guides inside a new biosphere reserve. In Sudan, auto tires are being recycled to make beds.

Five young innovators presented to a panel of seven judges from the private sector and the UN introduced several novel ideas: recycling 3-D printer plastic, storing digital data in living plants, and supporting sustainable farming with new phone apps. Reboot the Earth has designed bloc-chain software allowing a decentralized marketplace to prosper with sustainable development. In Finland, direct air capture technology attracts carbon dioxide from the air.

A spokesman from Denmark described how Global Alliance for Youth Climate currently has more than six hundred "ambassadors" educating their peers and involving themselves with decisionmakers in a number of countries. A student from NYU announced an Earth Uprising College aimed at integrating climate education into current courses. A Full Cycle Climate Fund accelerates the deployment of climate-restoring technologies.

A young man from Liberia said he'd come to "tell the adults it's about time they stop listening to their smart brains and start listening to their compassionate heart."

The day of Greta Thunberg's speech, Donald Trump showed up at the UN. The President popped in for fifteen minutes of the UN's day-long special session to address the climate crisis before leaving to give a talk elsewhere on religious freedom. As the president left the General Assembly hall, the Secret Service cordoned everyone near the entrance to step aside. Trump walked right past Greta, who simply stared at him. The clip of that moment quickly went viral on social media. That night, Trump felt compelled to paste a frame from her speech on Twitter, alongside his mocking summation: "She seems like a very happy young girl looking forward to a bright and wonderful future. So nice to see!"

Trump's snide dismissive remark wasn't deemed worthy of further reply. How dare he?

Washington's Climate Criminals:

Let's look at what's happened to our environmental laws and regulations since Trump assumed power, starting with the most recent outrage: At the dawn of 2020, the current administration announced a plan to severely weaken the fifty-year-old National Environmental Policy Act (NEPA) by eliminating climate change as a factor to consider in evaluating impacts of pipelines, highways, and other large-scale infrastructure projects. This could make moot a federal court ruling blocking construction of the Keystone XL fossil fuel pipeline, as climate change had not been properly taken into account.

A few months later came the COVID-19 lockdown, which the administration seized upon to brazenly gut more environmental protections: easing fuel-efficiency standards for new cars, freezing rules for soot air pollution, relaxing reporting requirements for polluters during the pandemic, continuing to lease public property to oil and gas companies, and advancing a proposal on mercury pollution from power plants that makes it easier for the government to claim that regulations are too costly to justify their benefits.

Trump's War on Science:

He's defunded the National Climate Assessment, compiled every four years, after the latest report predicted that climate change would bring catastrophic ecological and economic consequences to the US.

His new EPA-approved regulations are preventing experts from serving on congressional science boards, instead stocking them with industry researchers. It also proposed a rule that would restrict the use of science when developing health standards, if the underlying data is not "public and reproducible."

The attack on science permeates all agencies. The Interior Department reassigned its top climate scientist after he raised alarm about the impacts of climate change in July 2017. Lewis Ziska, a climate scientist with the Department of Agriculture, resigned last August after the USDA buried his latest research. "It's surreal," Ziska said of the Trump administration's censorship of climate science and research. "It feels like something out of a bad sci-fi movie."

Most publicly, and absurdly, after Trump made a false statement about Hurricane Dorian heading toward Alabama, his chief of staff pressured officials at the National Atmospheric and Oceanid Administration to release a statement supporting the President or face being fired.

Several of the fossil fuel barons named to the original Trump cabinet have since departed, but their replacements have been just as bad if not worse.

EPA Administrator:

One of Scott Pruitt's last acts was to announce that the EPA would no longer uphold a previous ban on using hydrofluorocarbons in appliances like refrigerators and air conditioning units. After Pruitt scurried back to Oklahoma while facing a dozen investigations for lavish spending of taxpayer money and using his position to enrich his family, a former coal lobbyist and unabashed climate denier quickly replaced him. Pruitt's replacement Andrew Wheeler, formerly Vice President of the Washington Coal Club, had presented Vice President Mike Pence an "action plan" for getting rid of the Clean Power Plan, withdrawing from the international climate accord to reduce emissions, eliminating federal tax credits for renewable energy, and cutting the EPA's work force in half.

In his first twelve months running the EPA, Wheeler killed or weakened dozens of safeguards with the sole intention of bolstering polluting industries' profit margins. As a result, according to the EPA's own accounting, millions of Americans will be breathing dirtier air and drinking filthier water. Wheeler eliminated the agency's Office of the Science Advisor, whose role was to counsel him on research which supported health and environmental standards. Some have suggested the EPA now stands for "Every Polluter's Ally."

Wheeler's most damaging move came when he signed a final rule to repeal and replace the Clean Power Plan that would have forced coal-fired plants to dramatically reduce their carbon emissions. Wheeler's "Affordable Clean Energy" edict gives states the authority to determine their own standards while setting no targets, which supplies them an option to do...absolutely nothing. At the same time, the EPA plans to gut the requirement for oil and gas companies to run inspections for methane leaks at their drilling sites. Over a ten-year time scale, methane is 85 times more powerful a greenhouse gas than carbon dioxide.

Last August, the EPA and Transportation Department rolled back an Obama-era rule to limit carbon emissions by requiring automakers to boost fuel economy to 54 miles per gallon by 2025 (in 2004, the average for cars and trucks was 24.6 miles per gallon). Wheeler and Transportation Secretary Elaine Chao first published an opinion piece in the *Wall Street Journal*, headlined "Make Cars Great Again," claiming that the existing standards "raised the cost and decreased the supply of newer, safer vehicles."

In fact, the Union of Concerned Scientists concluded that such a shift would cause another 2.2 billion metric tons of global warming emissions by 2040, and would cost drivers billions more to fill their tanks. California, joined by 23 other states, filed a lawsuit aimed at preserving their power to set stronger vehicle emission standards. Then, in a brazen abuse of federal power, Wheeler declared regulatory war on California— threatening to withhold billions in federal highway funds over the state's alleged failure to meet federal air-quality standards. Wheeler followed this up by demanding California come up with a "remedial" plan to address supposed water issues. Trump then accused the state of dumping needles into the ocean, with Wheeler raising the specter of "piles of human feces" from homeless camps polluting waterways.

Secretary of Interior:

Amid multiple probes into his shady real estate dealings and conduct in office (including a land deal with Halliburton in his Montana hometown), Ryan Zinke submitted his resignation in December 2018. By then, he'd overseen the biggest rollback of federal land protections in American history, and opened up nearly all of our coastal waters to offshore drilling.

Zinke's replacement is another pawn of the fossil fuel industry: his deputy, David Bernhardt, who previously represented a number of oil company clients that had business before the Interior Department. Since coming to the Interior, Bernhardt has spearheaded a review of the Endangered Species Act to make energy development easier on federal land and signed an order undoing agency rules for reducing environmental impacts, including best practices for mitigating climate change.

Secretary of State:

To this author's surprise, former ExxonMobil CEO Rex Tillerson (a featured "Horseman" in this book) had pushed Trump to keep "a seat at the table" in global climate change talks and remain part of the landmark Paris Agreement signed by 193 nations. Tillerson didn't succeed. After Trump fired him in May 2018, CIA Director and Koch brothers toady Mike Pompeo from Kansas moved in as his replacement.

Pompeo called the international climate accord a "feel good" deal that "didn't change a thing." And he's seemed thrilled about what's happening in the Arctic. "Steady reductions in sea ice are opening new passageways and new opportunities for trade," Pompeo said in May 2019. "This could potentially slash the time it takes to travel between Asia and the West by as much as twenty days."

He continued: "The Arctic is at the forefront of opportunity and abundance. It houses 13 percent of the world's undiscovered oil, 30 percent of its undiscovered gas, an abundance of uranium, rare earth minerals, gold, diamonds, and millions of square miles of untapped resources, fisheries galore."

Security of Energy:

Rick Perry's plan to provide new subsidies to coal and nuclear power plants was rejected by the Federal Energy Regulatory Commission (FEC) at a time when wind power actually generates more electricity than coal in his home state of Texas. Although the former governor had long questioned accepted climate science, Perry finally conceded during the sizzling summer of 2019 to an interviewer that "the climate is changing. Man, it's been changing forever. Where have you been?" He went on to muff his lines, or at least the English language, adding: "Are we part of the reason? Yeah, it is. I'll let people debate on who's the bigger problem here." Along with his tongue-tied response, Perry did go on to say that "it makes sense for us to have policies that reduce emissions. Common sense tells you, bring the cleaner burning fuels, bring the things that bring the emissions down." In October 2019, Perry turned in his resignation after being implicated in the Ukraine impeachment inquiry.

Secretary of Agriculture:

One man whose job appears to be safe is Sonny Perdue. As Governor of Georgia back in 2007, amid one of the state's worst droughts in decades, he'd led a prayer vigil on the Capital steps "to very reverently and respectfully pray up a storm." For the next two weeks, the epic dry streak worsened.

But Perdue didn't take the hint. Asked in an interview last June whether he believes climate change is caused by humans, he replied: "I think it's weather patterns, frankly. They change…It rained yesterday. It's a nice, pretty day today. The climate does change in short increments and in long increments." At a House Agriculture Committee meeting, Perdue joked about needing to give cows Pepto-Bismol to cut down on their flatulence, which emits tons of methane into the atmosphere.

Almost immediately after Trump took office, a USDA official in charge of soil health sent an email to senior staff urging they consider using the term "weather extremes" instead of climate change. An investigation by *Politico* determined that USDA routinely buries its own scientists' findings about the potential dangers of a warming world, including publicly burying a sweeping interagency plan for how to best respond.

The department has abandoned a "Climate Smart" effort launched under Obama, aimed at reducing agriculture's net emissions and sequestering over 120 million metric tons of carbon dioxide annually by 2025. USDA said at the time this would be equivalent to taking 25 million cars off the road, but that's now been scrubbed from their website.

The Impact of Big AG:

If you were asked to name the biggest contributor to climate change, you'd likely respond that it's the fossil fuel industry and corporate players like ExxonMobil. But that's not really true. In fact, it's a chemical behemoth called Monsanto, and a glyphosate-based herbicide known as Roundup, whose use on farmers' fields has soared 10,000 percent since 1974.

My co-author on this book, Robert F. Kennedy, Jr., is part of the legal team that won three major lawsuits against Monsanto over the past few years, after proving to juries that the company had long known Roundup and its active ingredient were probable human carcinogens and had been deliberately hiding this scientific fact. Their clients, suffering from non-Hodgkin's lymphoma, received millions in damages. Thousands more complaints are now being filed each month, and the number of plaintiffs lining up to sue has climbed past 18,400 as of this writing.

Since the Trump Justice Department approved Monsanto's sale to the German pharmaceutical giant Bayer in 2019, glyphosate has become the biggest headache aspirin can't cure. Such a migraine that Bayer's stock price has fallen around 33 percent since the deal closed, leaving its market value at $68 billion—barely above the $63 billion it paid to buy Monsanto. At the most recent shareholder meeting, investors took the unprecedented step of a no-confidence vote against Bayer's CEO Werner Baumann.

Glyphosate began as a broad-spectrum herbicide in the 1970s, meaning that it killed any vegetation it came in contact with. This limitation changed in 1996, when Monsanto introduced glyphosate-tolerant crops, starting with corn, then soybeans, cotton, and more. That meant farmers could apply it on and around their fields without impacting the crops. Global glyphosate use has risen almost fifteen-fold since then. By 2014, farmers sprayed almost a pound of Roundup on every acre of cropland in the US. The number one crop, field corn for livestock, is almost entirely sprayed with glyphosate.

When it comes to climate change, the UN's Intergovernmental Panel, titled "Climate Change and Land," specifies that urgent changes are needed in our food systems   It cites how Brazilian soybean farmers and cattle ranchers had torched the Amazon rainforest in order to make more space for their industrial-scale fields.

Since chemical agriculture ramped up worldwide forty years ago, we've lost one third of the earth's farmable topsoil. At the current rate, we'll run out altogether within sixty years. As happens with our bodies, glyphosate kills the microbes that break down organic matter in the soil. Healthy soils absorb water and carbon dioxide; damaged soils release water and carbon dioxide. Too much bare ground brings desertification. As soil turns to dust around the world, every year 40 million people are pushed off their land.

It's a vicious cycle. The global fertilizer market is worth $170 billion, spreading 44 billion pounds onto our crops every year. A recent study found that glyphosate reduces the nutrients in soil, creating a need for more synthetic fertilizers—which also kill soil microbes. Soil holds some 2.5 trillion tons of carbon, compared with 800 billion tons in the atmosphere and 560 billion tons in plant and animal life. The soil's capacity to absorb carbon is directly related to its health. Depleted soil cannot sequester carbon and allows fewer rain-making bacteria to be released into the air. By 2050, it's estimated that a billion people will be refugees of soil desertification.

The good news is that Roundup is falling out of favor. Costco has removed it from shelves in the US. Politicians from Austria to India are calling for bans on glyphosate. Belgium, Canada, and other countries have implemented restrictions on its use. In 2022, its authorization by the European Union will come up for review.

Soil Regeneration: Antidote to Climate Change:

A new documentary, *Kiss The Ground*, tells the story of the French Ministry of Agriculture presenting a pioneering plan at the 2015 UN Climate Summit in Paris. They called it the "Four for 1,000 Initiative"—a goal to increase the carbon content in the world's soils by 0.4 percent annually, which would sequester the same amount of carbon that humanity emits each year. This "drawdown," as it's called, would be achieved by shifting to agricultural practices that regenerate the soil by cutting back on toxic chemicals while reducing tillage and growing cover crops, as opposed to existing monocultures that require tons of fertilizers and herbicides.

It's not just fixing what's bad, but making things better. Regenerative agriculture has already been shown to grow more food per acre. Thirty countries at the Climate Summit signed a pledge to implement these methods—with three notable exceptions: India, China, and the US. Those countries happen to be the ones responsible for the majority of global CO_2 emissions.

But a shift is underway at the grassroots level, powered by agricultural activists whose innovations are both tried-and-true and cutting edge. Word is getting out through documentary films like *The Need to Grow*, produced by the Food Revolution Network, a documentary which examines the unique approach at a model urban microfarm in Irvine, California. Alegria Fresh is a zero-waste, solar-powered, one-acre operation that employs hydroponics and 100 percent organic soil-based growing systems. Founder Erik Cutter calls it "tractorless" farming, designed for man-made surfaces near where people live and work. Locally grown salad greens, herbs, and vegetables use 90 percent less energy and water, 70 percent less land, and half the fertilizer of conventional farms.

A 2018 study by the National Academy of Sciences estimated that as much as three billion tons of additional carbon dioxide could be captured and stored on global farmland if improved practices like adding manure or compost, changing to crops that add more carbon to the soil, or planting cover crops in the off season that then break down to organic matter.

Toronto native Rachel Parent is the founder of several organizations in Canada seeking to educate young people about environmentally conscious food production. The nineteen-year-old serves as Youth Director for Regeneration International, which is building alliances in the American Midwest, Mexico, India, Guatemala, and Belize. Parent wrote in 2019:

> Whereas the current industrial model has narrowed the genetic base and makes the entire food system vulnerable to climate change, regenerative agriculture using agroecological approaches lead to more crop diversity and enhanced food security.
>
> We can no longer tinker at the edges of a fundamentally flawed approach to economic activity, food systems and soil management that relies on fossil energy and disrupts ecosystems. Regenerative agriculture has the potential to solve our growing climate crisis, bring together communities, create healthy economies and protect the very soil we stand on. The solution is right beneath our feet.

The USDA's latest census shows that the number of farmers under age thirty-five is on the upswing. Remarkably, 69 percent of those surveyed have college degrees. A National Young Farmers' Coalition maintains that this surge of first-time growers is driving a potential sea change in the industry. They are capitalizing on the demand for local food by selling directly to consumers through a wide diversity of crops and livestock. Seventy-five percent describe their practices as "sustainable," and 63 percent as "organic." They're moving into communities and neighborhoods focused around gardens and green spaces.

"Young farmers are on the front lines of climate change," the Coalition report states, "experiencing unpredictable weather, severe storms, drought, pests, and disease. Congress should prioritize climate-smart conservation programs that promote soil health and resilience, increase beginning farmer access to these programs, and provide adequate staffing and technical assistance."

Technological Fixes: What Works and What Doesn't:

Geo-engineering is the large-scale modification of our earth systems to alleviate climate change. But most such approaches aren't going to cut it—and could even make things worse. For example, the idea of spraying aerosols into the atmosphere would theoretically cool the planet by reflecting sunlight away from the earth. But, on the other hand, blocking part of the sun from space would also reduce the energy available for solar power and even photosynthesis. There's also talk of spraying seawater into the atmosphere to achieve the same end, by increasing cloud reflectivity and condensation. And of putting trillions of miniscule solar reflectors out in space. Or releasing tiny sunlight-reflecting particles that mimic a major volcanic eruption. Other geo-engineering schemes include ocean fertilization, where extensive areas get sprinkled with iron or other nutrients aimed at artificially increasing growth of phytoplankton that soak up CO_2. A de-facto UN moratorium has been in place since 2008, because this approach would trigger harmful algal blooms.

In 2018, the amount of CO_2 estimated to have entered the atmosphere from fossil fuels hit around 37 billion tons. Here's the rub: we not only need to cut emissions radically, but some climate experts say that, by 2050, we must remove as much as 1,500 billion tons of carbon dioxide that's already been emitted.

The majority of environmentalists argue that we don't need to artificially change nature, but rather improve nature conservation. Most carbon capture technologies have been developed for point sources like coal-fired power plants, but that doesn't address the emissions coming out of cars and homes. How to achieve a better "fix" assumes ramping up a technology known as BECCS—bioenergy with carbon capture and storage. As Matt Frost has written in *The New Atlantis* (December 2016), "with this technology, plants are grown and then burned for energy or processed into fuels like ethanol. Carbon emissions from power plants that burn biomass are captured and stored in permanent reservoirs. Because plants use carbon dioxide to grow, bioenergy with carbon capture and storage has 'negative emissions'—meaning that it actually draws more carbon dioxide out of the atmosphere than it releases."

Although most gasoline sold in the US contains a small amount of corn-derived ethanol, currently only five operating BECCS projects exist in the world and these store one-and-a-half million tons of CO_2 per year. An IPCC report estimates that the technology will need to increase that storage to seven *billion* tons annually by 2050 to meet the 1.5-degree target. "No proposed technology is close to deployment at scale," the IPCC admits. And we'd need to set aside as many as three million square miles of farmland for bioenergy crops, which is almost as large as the contiguous United States.

However, physicist Klaus Lackner, director of the Center for Negative Carbon Emissions at Arizona State University, has a more feasible idea. He sees dumping CO_2 into the atmosphere as a form of littering, constituting a waste management problem above all else. Just as we learned to build sewer systems and collect household garbage, so can existing technologies for CO_2 capture and disposal (storage) be utilized. Rather than relying on biological methods, Lackner proposes a chemical engineering approach for Direct Air Capture of CO_2 from the atmosphere, which is then stored in subsurface rock formations.

In April 2019, Arizona State University and Silicon Kingdom Holdings announced a joint venture to deploy Lackner's proprietary technology, which removes CO_2 without needing to draw air through the system mechanically using energy-intensive devices. Instead, wind blows air through the system onto synthetic "mechanical trees," collectors to which carbon dioxide sticks a thousand times faster than from natural trees of similar size. The CO_2 is then recycled by either burying it or be sold for re-use in synthetic fuels and other applications.

Lackner has written: "In the waste management paradigm, you simply pay to remove your own emissions from the atmosphere, just as you pay to have your sewage processed....For example, a city could run its own carbon disposal site, or an oil company could offer carbon-neutral fuels at the pump....Imagine a button at the gasoline pump where individuals can choose to pay to have the 20 pounds of carbon dioxide that are released from a gallon of gasoline recovered and properly disposed of."

The coal and oil industries receive an astounding $5.2 trillion annually in subsidies. Take those away and they would be unable to compete in the market. In fact, last year 63 percent of the new generation constructed worldwide was renewable. The cheapest form of energy today, according to Bloomberg Energy, is solar at about 17 cents per kilowatt hour—whereas nuclear is ten times that, coal is about seven or eight times that, and natural gas is triple that. If people really believed in free market capitalism, we could solve this problem overnight.

A free market-based economy is supposed to reward good behavior, which is efficiency, and punish bad behavior, which is inefficiency and waste. We need rules to rationalize our marketplace to create a society we're all proud of and that will sustain future generations.

The free market is the most powerful economic engine ever devised by humanity. But it has to be harnessed with a social purpose, otherwise it will just lead us down the inevitable road of political oligarchy and corporate kleptocracy and environmental apocalypse.

Youth Solutions, 2020:

Earth Guardians trains diverse youth to be effective leaders in the environmental, climate, and social justice movements. Through the power of art, music, storytelling, civic engagement, and legal action, they're creating impactful solutions to some of the most critical issues we face as a global community. www.earthguardians.org

Earth Uprising is a global, youth-led organization focusing on climate education, climate advocacy, and mobilizing young people to take direct action for their future. www.earthuprising.org

Extinction Rebellion Youth is led by a community of young people within Extinction Rebellion, a network focused on persuading governments to act on the climate and ecological emergency. www.xryouth.org

Fridays for Future USA is a people-led movement around the climate crisis. Founded in August 2018, Fridays for Future was inspired by teen activist Greta Thunberg, who sat in front of the Swedish parliament every school day for three weeks to protest against the lack of action on the climate crisis. www.fridaysforfuture.org

Future Coalition is a national network and community for youth-led organizations and leaders. The Future Coalition works collaboratively to share resources and ideas, all with a common goal of making the future a better, safer, and more just place for everyone to live. www.futurecoalition.org

International Indigenous Youth Council seeks to organize youth through education, spiritual practices, and civic engagement to create positive change in our communities. Through action and ceremony, the IIYC commits to building a sustainable future for the next seven generations. www.indigenousyouth.org

Kids Right to Know, founded by nineteen-year-old Toronto native Rachel Parent, is dedicated to educating youth about environmental justice and food safety, with a focus on regenerative agriculture that restores carbon to the soil. www.kidsrighttoknow.org

See also: "17 organizations promoting regenerative agriculture around the world," at www.foodtank.com

Sunrise Movement is a youth-led movement of young people committed to stopping the climate crisis. Sunrise Movement is currently leading actions around a Green New Deal and need for a Democratic debate dedicated to climate change. www.sunrisemovement.org

US Youth Climate Strike is a youth-led movement that helped organize over 424 student strikes occurring in at least 45 states on March 15, 2019. www.youthclimatestrikeus.org

Zero Hour is an intersectional movement around climate change. In 2018, Zero Hour organized the first official Youth Climate March in 25 cities around the world and laid the groundwork for the climate strike movement. In July 2019, Zero Hour hosted the Youth Climate Summit, a weekend-long event featuring 350 attendees from across the world participating in workshops and programs to enhance their advocacy in the fight for climate justice. www.thisiszerohour.org

Data collected from www.climatesolutions.org

Foreword

By David Talbot

Their names will go down in infamy. They are the dark lords of the energy underworld—the oil, gas, and coal executives who continue to exhume fossil fuels from the earth as the planet grows hotter by the year.

Some of their names are already infamous. And some are being made more so by their intimate association with the Trump presidency—such as fossil fuel titans Charles and David Koch, who rule the world of political dark money; Secretary of State and former ExxonMobil CEO Rex Tillerson; and Environmental Protection Agency chief Scott Pruitt, who as Oklahoma attorney general fended off efforts to regulate widespread fracking in the state, which has led to a surge of earthquakes and pollution.

Others have pursued their dirty wealth relatively discreetly, away from the public spotlight, such as former Peabody Coal chief Greg Boyce and Oklahoma fracking king Harold Hamm. But their impact on the health of the planet has been no less destructive.

These are the horsemen of the apocalypse whom author Dick Russell investigates and indicts in the following pages. These are the men whose colossal fortunes and power come at humanity's expense, as the earth burns.

The growing calamity of climate change is usually presented in scientific and environmental terms that are either too abstract or too starkly real for average citizens to know how to respond. But the alarming changes in weather patterns—the droughts, firestorms, flooding, and freak storms that are wreaking havoc around the world—can be traced to the business and political decisions made by the men profiled in this book.

These men like to think of themselves as not only generators of great wealth and economic progress but pillars of their community. But this book shows them for what they truly are. As author Dick Russell writes, "These dark lords like to pose as good family men, benefactors of charities and the arts, upstanding pillars of their community. But first and foremost they are enemies of life on earth. This book has sought to put a face to the entrenched evil that has pushed us to the point of no return."

There is nothing abstract about the energy policies pursued by these men—they are directly linked to the heating of the planet. These energy moguls and power brokers are guilty of crimes against nature. And, as with any criminals, they must ultimately be held accountable for their crimes. In the current political climate, of course, this is impossible—at least in Congress or the courts. But if we care about the lives of our children and grandchildren, we must finally put a stop to these men's reckless profiteering and plundering.

These men claim they care about future generations on earth. Tillerson even served as president of the Boy Scouts of America. But at the same time that he was devoting himself to the physical and moral health of young men, Tillerson and fellow Exxon executives were covering up frightening evidence produced by their own company scientists about fossil fuels' impact on planet warming. There is a special ring in hell reserved for men such as this—those who put their prosperity and success ahead of humanity, including their own flesh and blood.

Despite the grim prospects we face in the age of Trump, there is reason for hope, notes Russell. Some of those whose wealth is based on Big Oil not only have awakened to the existential threat of climate change but are doing something significant about it. The dynasty built on the enormous fortune of John D. Rockefeller, America's original oil tycoon, has begun to disinvest from fossil fuel companies like ExxonMobil. "We all have a moral obligation," stated Valerie Rockefeller Wayne, the chair of the Rockefeller Brothers Fund. "Our family in particular—[since] our lifestyles come from dirty fossil fuel sources."

Although the Trump administration is stacked with horsemen of the apocalypse, the movement to shift the planet from dirty to clean energy sources is growing stronger. Even school-age children—insisting they have a right to a safe and healthy future—are now taking climate change polluters to court.

As Russell concludes, we are engaged in an epic struggle between the forces of greed and devastation on one side and those of life. By identifying the enemies of life and exposing the ways they operate, *Horsemen of the Apocalypse* strikes a blow for all those who are fighting for the future.

David Talbot
March 2017

Introduction to 2017 Edition

By Robert F. Kennedy, Jr.

Not long ago, the legendary economist Amory Lovins showed me two photos, taken ten years apart, of the New York City Easter Parade. A 1903 shot looking north from midtown showed Fifth Avenue crowded with a hundred horse and buggies and a solitary automobile. The second, taken in 1913 from a similar vantage on the same street, depicted a traffic jam of automobiles and a single lonely horse and buggy.

That momentous shift occurred because, over a thirteen-year period, Henry Ford dropped the nominal price of the Model-T by 62 percent. While wealthy New Yorkers led the transition, the remainder of America quickly followed. Between 1918 and 1929, according to Stanford University lecturer Tony Seba, American car ownership rocketed from 8 percent of Americans to 80 percent—because DuPont and General Motors devised a financial innovation called car loans, which soon accounted for three quarters of auto purchases. The buggy drivers never saw it coming.

Compare that platform for disruption to the economic fundamentals of today's solar industry. Over the past five years, photovoltaic module prices have dropped 80 percent, and analogous home solar financing innovations have spread like wildfire. Three-quarters of California's rooftop solar has been innovatively financed, with no money down, including the system I installed on my own home. NRG Solar leased me a rooftop solar array with zero cost to myself and a guaranteed 60 percent drop in my energy bills for twenty years. Who wouldn't take that deal? And solar costs continue to fall every day.

Dramatic drops in labor costs and improvements in construction practices have cut the cost of utility-scale solar plants to around $1 billion a gigawatt. Compare this to the $3 to $5 billion per gigawatt cost of constructing a new coal or gas plant, and the $6 to $9 billion per gigawatt cost for a nuclear plant. We can make energy by burning prime rib if we choose to, but any rational utility seeking the cheapest, safest form of energy is going to choose wind or solar. That's why, according to the Federal Energy Regulatory Commission, in the first eleven months of 2016 renewables constituted over 50 percent of newly installed electrical generation capacity—surpassing natural gas, nuclear power, coal, and oil combined. Let's face facts. The carbon incumbents are looking at their own imminent apocalypse.

And the real savings for solar and wind comes at the back end— ZERO FUEL COSTS! Unlimited photons rain down on the earth every day for free. Transitioning to clean fuel only requires that we build the infrastructure to harvest and distribute the photons. That infrastructure will bless America with a magical promised era of "free fuel forever."

Internal combustion engines are racing toward the same kind of apocalyptic disruption as the horse and buggy. According to calculations by John Walker of the Rocky Mountain Institute, the current operating cost of an electric car is about one-tenth the cost of an internal combustion engine. The range and performance of EVs now exceeds those of traditional gasoline cars. That's why the world's fifteen top auto companies all launched new EVs in 2015. If you believe in free markets, then the day of the internal combustion engine is over.

The markets have already seen the future. The top fifty coal companies are now either in Chapter 11 bankruptcy or on the brink. The three largest coal companies—Arch, Consol, and Peabody—have lost 80 percent of their value over the past two years. Looking at these landscapes, Lovins remarked to me, "The meteor has hit. The dinosaurs are doomed. It's just that some of them are still walking around causing trouble."

With these rich indices of imminent change, America, prior to the 2016 election, was on the verge of leading the global transformation away from destructive reliance on the dirtiest, filthiest, poisonous, addictive, war-mongering fuels from hell, to a sunny new age of innovation and entrepreneurship, of abundant and dignified jobs, of a democratized energy system and widespread wealth creation, powered by the clean, green, healthy, wholesome, and patriotic fuels from heaven.

Renewable energy sources such as wind and solar create high-paying jobs, promote small businesses, create wealth, democratize our energy sector, give us local, resilient power, and reduce dependence on foreign carbon. They are therefore good for the economy, good for our national security, and good for democracy and our country.

And every American will benefit from the cornucopia of economic and political bounties that accompany a decarbonized nation—no more poisoned air and water, but clean rivers and bountiful oceans, with fish that are safe to eat. No more exploded mountain ranges. No more crippling oil spills in the Gulf, in Alaska, or in Santa Barbara. No more worries about acid rain deforesting our purple mountains' majesty and sterilizing our lakes. No more fretting about acidified oceans destroying our coral reefs and collapsing global food chains and fisheries. No more ozone and particulate pollution sickening and killing millions of our citizens. No more damaged crops and corroding buildings. No more tyrannical petro-states subjugating their peoples and victimizing their neighbors. And no more oil wars.

While enticing to most Americans and consistent with the historical idealism of an exemplary nation, this portrait of the future represents a fearful nightmare for a certain segment of our population—a segment that is willing to mount all-out civil war to prevent it from happening, an apocalyptic war that threatens to sacrifice the planet.

And we *are* engaged, as Abraham Lincoln declared, "In a great Civil War." In the 1860s, it took a bloody civil war for America to transition away from an archaic and immoral energy system—one dependent on free human labor. In 1865, the entrenched interests who profited from that system were willing to sacrifice our country, and half a million lives, to maintain their profits.

This time, instead of a slave-holding gentry, the entrenched defenders of the system are the carbon tycoons described by Dick Russell in *Horsemen of the Apocalypse*. These are the apocalyptic forces of ignorance and greed that are out to liquidate our planet for cash. Russell shows that, to the extent they have a moral compass, it's pointed straight at hell. Like the Horsemen from the Book of Revelation, these actions are herding humanity toward a dystopian nightmare of their creation. The archetypal Horsemen are David and Charles Koch, whom you will meet in Chapter 8. Koch Industries, you will learn, is not a benign corporation. It's the template of "disaster capitalism," the command center of an organized scheme to undermine democracy and impose a corporate kleptocracy that will allow greedy billionaires to cash in on mass extinction in our biosphere and the end of civilization. To the Koch brothers, the renewable revolution is their personal apocalypse that must be averted at all costs.

With their industry bereft of its economic rationale, the only way the carbon incumbents can maintain their economic dominance is by deploying their wealth and political power to subdue the market forces, to delay and derail cheap efficient renewables, and to impose a continued dependence on expensive and inefficient oil and coal through massive economic interventions managed by their political toadies. The Koch brothers have become the masterminds of this strategy. They sit at the apex of the richest industry in the history of the planet and control the largest privately owned oil company on Earth. Their strategic advantage in the battle over the future of our energy system includes their enormous personal wealth and the wealth and power of the companies they control. Their carefully cultivated political connections and, above all, societal inertia, fortified by $23 trillion of carbon infrastructure, has impeded America's transition to a new energy economy. While owned by the industry, that infrastructure, ironically, was primarily paid for by taxpayers. These form the carbon cartel's principal arsenals in the great civil war.

As economic forecasts for the industry have grown increasingly dire over the past decade, the carbon cartel has moved frantically to build more infrastructure including LNG facilities, refineries, coal and oil export terminals, and rail terminals, in order to bind up America in 16,000 miles of new pipeline that will further shackle our country, ironbound, to an archaic and destructive fossil fuel economy long after any economic rationale for coal, oil, or gas has expired. The infrastructure strategy effectively recruits bankers, pension funds, and Wall Street financial houses to the side of antiquated carbon in this civil war. The only hope for those investors to recoup their investments is if oil flows continuously through those pipelines for the next thirty years.

Dick Russell largely completed this book a month before the 2016 presidential election. Four weeks later, to his great surprise, the Horsemen described in these pages assumed the pinnacle of power. With the central goal of preserving their fossil fuel profits, they guided an inexperienced president on a course that rapidly collapsed the foundations of America's moral authority and idealism, and fundamentally altered the relationship between America and the world—including our reputation as a global force for good. Their reigning foreign policy posture was an indifference to America's traditional concerns with justice, democracy, and climate; to our historic skepticism toward tyrants; and to those traditional alliances that have promoted global stability since World War II. Domestically, Trump's advisors turned their attention to dismantling the science safety net and commoditizing and monetizing every aspect of human discourse—adopting policies that will amplify the wealth of billionaires, even as they sicken our citizens and destroy our planet. The Book of Revelation described the Four Horsemen as War, Conquest, Pestilence, and Death. Donald Trump's choice to invite a group of conscienceless oil men to govern the country has brought such chilling metaphors to the foreground as more than an obscure biblical reference.

In the summer of 2016, it seemed that a convoy of clown cars was transporting Donald Trump in what would become his unlikely blitzkrieg toward the GOP nomination. I was oddly relieved. Like other Americans, I believed that Donald Trump would be an easy candidate to stop in the November general election. More importantly, Trump didn't seem as purposefully malicious toward the future of the planet as were his principal rivals, Ted Cruz, Marco Rubio, Scott Walker, and Rick Perry. I had known Donald Trump for many years. I had successfully sued to block him from building two golf courses in the New York upstate reservoir watershed. I knew he was no friend of the environment, but neither did he appear to be ideologically hidebound to a pro-pollution worldview. Indeed, he seemed less shackled to dogma, or obligated by encumbrances than any other Republican presidential candidate. He had no obvious fealty to the oil industry. Alone among the seventeen rivals for the Republican nomination, Trump had never taken money from the oil and gas tycoons. Most comforting, there seemed to be a deep gulf of enmity between Trump and the billionaire Koch brothers, the undisputed leaders of Russell's *Horsemen of the Apocalypse*. Taken together, Charles and David Koch, with $48 billion apiece, are the richest men on Earth, according to *Forbes'* latest list. (Bill Gates has $86 billion). The siblings' father, Fred Koch, had made a fortune building refineries for Hitler and Stalin and used his money to cofound the racist John Birch Society. The boys, Charles and David, have deployed their oil-and-gas fortune to bankroll an array of think-tanks and politicians opposing clean energy and remedial action on climate change.

Teddy Roosevelt observed that American democracy could never be destroyed by a foreign foe. But he warned that our defining democratic institutions would be subverted from within by "malefactors of great wealth."

Because of their singular focus and limitless wealth, I considered David and Charles Koch, rather than this orange-haired GOP candidate, the greatest threat to American democracy. Politics is driven by both money and political intensity. While they have plenty of money, the Koch brothers' policy agenda—tax breaks for the rich, unregulated pollution, and permanent national reliance on dirty fuels—does not make an attractive vessel for populism. In order to recruit ground troops, the Koch brothers have made themselves wizards in the alchemy of demagoguery, wielding evangelical religion, dog-whistle race baiting, and patriotism as flypaper to their cause. They have built extensive organizations to engineer a hostile takeover of our democracy by polluting corporations. In her book *Dark Money*, Jane Mayer shows how the two oil men conceived and funded the Tea Party movement, which hijacked the Republican Party and drove it to the far right. In order to consolidate power over the past two decades, they worked out and financed a methodical project to take over state legislatures. Their lucre and organizing machine has helped to give right-wing Republicans control of sixty-seven of ninety-eight legislatures—the bodies that draw up electoral districts. With those levers in hand, their lackeys in the various state capitals use gerrymandering, voter fraud, voter ID laws, and mass voter purges to engineer permanent Republican majorities on the state and federal level. Their Tea Party movement took over the US Congress and blocked Obama's environmental agenda, with the resilience fortified by their control of the statehouses. But the biggest electoral prize the Koch brothers had yet to achieve was to have their candidate take the White House, with the power to populate and dismantle the agencies, primarily Energy, Interior, and EPA, that regulate—and bedevil—the Kochs' industries. They had many willing errand boys among the Republican presidential field, and, with one notable orange exception, just about all of the GOP candidates had made the pilgrimage to Wichita to genuflect and kiss the rings at Koch headquarters.

The Koch brothers claim, in their rhetoric, to embrace a theology of free market capitalism. But if you look at their feet instead of listening to the seductive noises that issue from their mouths, or the glossy pronouncements of their phony think tanks, the truth is clear: these men despise free markets. Instead, they advocate for a system of cushy socialism for the rich, and a savage, merciless, dystopian capitalism for the poor. The real purpose of the "think tanks" they created and fund—such as the Heritage Foundation and the Cato Institute—is not to promote free market capitalism but to gin up the philosophical underpinnings for a scheme of unrestrained corporate profit taking and a destructive national addiction to carbon-based fuels upon which their fortunes rely.

As discussed earlier, new renewable technologies are now so efficient that wind and solar generation and electric cars are beating their carbon-based competitors, even in the rigged markets and on slanted playing fields. Carbon's economic model is looking at the same bleak future the horse and buggy industry faced in 1903. So what do you do when your profits rely on a fading economic model? If you are the unscrupulous Koch brothers, you deploy your money as campaign contributions—a legalized form of bribery—to get your hooks into a public official who will allow you to privatize the commons, dismantle the marketplace, and rig the rules to give you monopoly control. Renewable energy sources and free markets pose an existential threat to the Koch's business model. So the Kochs have deployed their front group, ALEC—the American Legislative Exchange Council—in every state, working with local legislators to create public subsidies for oil infrastructure and to weaken support for wind and solar. The Koch brothers' purpose in purchasing our political system is to engineer monumental subsidies and market failure, which are their formulae for profit.

The Kochs' political ascendancy was facilitated by another oilman, George W. Bush. A decade ago, I wrote a bestselling book, *Crimes Against Nature*, detailing Bush's war against the environment. As is almost always the case, environmental catastrophe was preceded, in the Bush debacle, by the subversion of democracy. Bush landed in the White House after a stacked Supreme Court, dominated by his father's appointees, issued a partisan 5–4 decision, freezing the 2000 election Florida vote recount that would have shown Bush losing both the popular vote and the Electoral College. Bush thereby stole the presidency from Senator Al Gore, the greenest presidential candidate in our history. That decision turned the White House over to two Texas oilmen, Bush and his vice president, Dick Cheney, who was the CEO of oil service company Halliburton and the owner of millions of dollars of Halliburton stock, which would appreciate enormously during Cheney's administration. Seventeen of the top twenty-one people in the new administration hailed from the oil patch or allied industries. Bush's secretary of state, Condoleezza Rice, was on the board of Chevron, which named an oil tanker after her. Bush's ascendancy to the presidency was the beginning of a hostile takeover of our government by the oil industry, which would finally be completed by President Trump. Transforming America into a petro-state was not just bad for the environment; it was a disaster for American democracy. Cheney immediately convened ninety days of secret meetings with carbon and nuclear industry CEOs, during which he invited the nation's worst polluters to rewrite environmental laws to make it easy to drill, to burn, to extract, to frack, to ship, and to distribute carbon fuel. It was an all-out victory for the carbon industry and an unconditional defeat for humanity. Even as they dismantled America's environmental laws, Bush and Cheney stocked the regulatory agencies with industry lackeys and profiteering cronies who weakened and auctioned off America's public lands and forests to their campaign contributors, at fire sale prices.

The oil and coal industries are, by nature, authoritarian. A nation's political system generally reflects the economic organization of its principal industries. In a dynamic known as the "resource curse," nations dominated by carbon industries customarily tilt toward autocracy and away from democracy. When oil money merges with political power, the outcome is almost always the same: yawning gaps between rich and poor; the expansion of military, police, and intelligence apparatuses; the diminution of civil and human rights; the disappearance of transparency and public participation in government; the expanded popularity of torture, detention, and eavesdropping; the use of nationalistic propaganda and deceit to win elections and to justify unpopular politics; and an aggressive, bellicose, and imperialistic foreign policy. Under George W. Bush, the carbon cronies quickly bent US foreign and domestic policies to serve Big Oil's bottom line.

The White House lied America into an oil war that killed a million Iraqis and almost 4,500 US soldiers, and maimed tens of thousands more. Bush and Cheney cut taxes on the wealthy and charged their $4.3 trillion war on a credit card for our children to pay. Meanwhile, we lost eight critical years in the battle to avert the most catastrophic impacts of climate change. The oilmen who helped bring Bush to power profited like princes from his policies. As a gesture of gratitude to our country, the oil companies raised gasoline prices and watched company profits soar to historic highs. Patriotism is great—so long as it pays!

But perhaps the most grievous wound to American democracy from the mayhem of the Bush presidency was the Supreme Court decision in *Citizens United v. Federal Election Commission*. Bush appointed two right-wing corporatist US Supreme Court Justices—John G. Roberts in 2005 and Samuel Alito in 2006. These men were not traditional conservatives. The only consistent thread running through their judicial decisions was the consistent elevation of corporate power. The *Citizens United* decision was the most sweeping expansion of corporate power in this century. That case effectively overruled a century of corporate campaign finance restrictions that limited a corporation's ability to purchase federal political candidates. *Citizens United* unleashed a tsunami of corporate cash in the 2010 and 2012 election cycles, when an estimated $2 billion was spent in the race to capture the White House. The nearly $760 million that the Koch brothers, alone, put into the national elections in 2016 is comparable to the total amount spent, historically, by either political party. Their campaign organization was nearly as formidable as the Republican Party, with 1,200 election operatives. Election data show that in 95 percent of federal elections, the candidate with the most money wins. So democracy is for sale and, predictably, the rich are buying themselves politicians and then deploying them to reduce taxes on their class and to rid themselves of pesky regulations that protect public health and the common environment. Under this new rubric, the representatives and senators who dominate Congress can no longer be thought of as public servants. They are the indentured servants of the Koch brothers and their ilk, engaged in the mercenary enterprise of ransacking America and humanity on behalf of Big Oil. America has transitioned from the world's model democracy to a corporate kleptocracy.

So the Kochs' open disavowal of Donald Trump during the presidential campaign was a comfort to me. Charles Koch compared the Trump-Hillary race to a choice "between cancer and a heart attack." Trump, in turn, derided his Republican rivals as Koch "puppets." I was relieved by the wide and hostile rift between Trump and the Kochs. Perhaps, I thought, America was safe, for the moment, from the existential threat of having oil tycoons again control our government. So it was breathtaking how quickly president-elect Trump pivoted against the populism that he rode to power, and into the welcoming arms of Wall Street robber barons, oil patch tycoons, and flat earth oligarchs he had vilified along the way, including the despised Koch brothers. Even before inauguration day, he got busy turning our government over to the apocalyptic forces of ignorance and greed.

Climate Change

Consolidation of power by the oil and coal barons began immediately after the election; president-elect Trump's transition advisors emerged as an oil industry dream team. Despite the initial antipathy between Trump and the Koch brothers, once he secured the nomination, Donald Trump extended the olive branch to the flat earth oligarchs from Kansas. His choice of Indiana Governor Mike Pence as running mate was the first ominous sign that the rift had healed. Governor Pence had financed his political career with a steady flow of Koch cash and had demonstrated his fealty to the Kochs by hiring Marc Short as his gubernatorial chief of staff. Short had previously been president of Freedom Partners, the Kochs' political arm. As governor, Pence made Indiana a proving ground for the radical right-wing experiment in corporate domination devised by Koch-funded think-tanks.

Three days after the 2016 election, Pence displaced New Jersey Governor Chris Christie to become Trump's overseer of the various agency transition teams. By that time, the writing was on the wall, and the penmanship was that of David and Charles Koch. David Koch attended Trump's election night celebration. Trump soon appointed Marc Short as his director of legislative affairs, and stocked his transition team with Koch organization veterans, such as Tom Pyle, Darin Selnick, and Alan Cobb, and transition team executive committee members, Rebekah Mercer and Anthony Scaramucci. According to *The Wall Street Journal*, an astonishing 30 to 40 percent of Trump's advisors had Koch pedigrees. These were the men and women who would shape the new president's agenda.

Trump appointed a notorious Koch toady, Myron Ebell, to supervise his EPA transition. I've watched Ebell's antics for decades. He is a professional deceiver. Ebell served as director of the Center for Energy and Environment at the Competitive Enterprise Institute, a Washington think-tank formerly funded by ExxonMobil and the Kochs, and staffed primarily by "experts" and operatives, lately employed by Koch Industries and the Koch's web of shadowy non-profit oil industry advocacy groups. Ebell, once a staunch global warming denier, has recently retrenched; "Yes, we are causing climate change," he now admits, "but it's a good thing." Ebell preaches that the "mild global warming that has occurred since the end of the Little Ice Age in the mid-nineteenth century has been largely beneficial for humanity and the biosphere. Earth is greening, food production has soared, and human longevity has increased dramatically."

Ebell's seven-person team included David Schnare, a lawyer who spent thirty-three years at the EPA before matriculating to institutes funded by the Kochs. Schnare made his bones as a polluter's shill by filing legal actions demanding to inspect the email inboxes of EPA administrators and climate scientists. In Trump's new era of "alternative facts," there was no one better suited to purge the agency of credulous climate change believers.

Steve Groves led the State Department's "landing team." Groves, a policy wonk at the Koch- and Exxon-funded Heritage Foundation, wrote a post-election article calling for the United States to pull out of the United Nations Framework Convention on Climate Change as a prelude to refuting the Paris Agreement.

The Department of Interior transition fell under the leadership of Doug Domenech, director of the Fueling Freedom Project for the Koch-funded Texas Public Policy Foundation. That group's mission is to "explain the forgotten moral case for fossil fuels." Domenech knows how to make the system work for industry; during George W. Bush's presidency, he served as White House liaison and deputy chief of staff at the Interior Department, facilitating Bush's efforts to turn federal lands over to oil, gas, and mining interests and to timber barons.

President Trump's transition overseer at the Department of Energy was Michael Catanzaro, a registered Koch Industries lobbyist. His successor is Thomas Pyle, former president of the Institute for Energy Research, a think-tank founded by Charles Koch. Before joining that chamber for charlatans, Pyle was Koch Industries' director of federal affairs. Pyle is also president of the American Energy Alliance, another fossil fuel front group that receives a pipeline of cash from Koch, Exxon-Mobil, and Peabody Energy. (You'll learn much more about the Peabody CEO in Chapter 7 of this book.)

Pyle mapped out "a big change" in an email to supporters in mid-November. He promised a "100-day plan" and a "200-day plan" to roll back America's clean water and climate change protections. America, he promised, will pull out of the Paris Climate Agreement, and the EPA will jettison the dreaded "social cost of carbon" algorithm used to calculate the costs and benefits of climate change.

In December, eight hundred US scientists and energy experts sent a letter to president-elect Trump asking that he publicly identify global warming as a "human caused, urgent threat." They went on: "If not, you will become the only government leader in the world to deny climate science. Your position will be at odds with virtually all climate scientists, most economists, military experts, fossil fuel companies and other business leaders, and the two-thirds of Americans worried about this issue." Trump answered this urgent plea by the world's most highly credentialed climate scientists during a Fox News interview in mid-December, assuring the audience that "nobody really knows" whether climate change is real. He said he was "studying" whether to pull America out of the Paris Climate Agreement, the hard-won treaty to reduce greenhouse gas emissions that has been signed by 196 countries. There is little doubt about who is providing him crib-notes.

The ominous direction toward global catastrophe crystallized as Trump announced his cabinet and other key positions.

SECRETARY OF STATE: REX TILLERSON

"And I looked, and behold a pale horse; and his name that sat on him was Death, and Hell followed with him. And power was given unto them over the fourth part of the earth, to kill with sword, and with hunger, and with death, and with the beasts of the earth."

—*Revelation*, 6:8.

In a breathtaking act of supplication to Big Oil, the new president gave his first cabinet appointment to Russell's first Horseman, ExxonMobil CEO Rex Tillerson. Tillerson has never been mistaken for an American patriot. As Exxon CEO, he often adopted company policies that were contrary to US interests, including a lucrative deal with Russia to drill in the Arctic. When a shareholder asked Tillerson's predecessor and mentor, Lee Raymond, whether the company should be improving US refinery capacity as a matter of national security, Raymond dismissed patriotism as an absurd distraction from profits. He famously declared,

"Exxon is not a US company." Tillerson's worldview is dictated by his forty years of service to the selfish ideologies of a corporation that is locked in a ruinous battle against humanity and American values.

Trump's critics wondered whether his peculiar choice to hand US foreign policy over to the world's most visible and notorious oil man was a favor to Russian dictator Vladimir Putin. As Exxon chief, Tillerson put aside scruples to align Exxon with the bloodthirsty tyrant, a choice that made Tillerson Putin's favorite American businessman. In 2013, Vladimir Putin personally presented Tillerson with Russia's ultimate honor to a foreigner, the Order of Friendship Award, after Tillerson signed controversial deals with the state-owned Russian oil company. In 2011, Tillerson flew to Russia to sign a $500-billion arrangement to jointly drill in the Arctic Shelf and the Black Sea and to develop shale oil in Siberia. Tillerson's company allegedly lost around $1 billion dollars due to sanctions the Obama administration placed on Russia after Putin annexed the Crimean Peninsula.

Tillerson responded by directing ExxonMobil's PAC to donate $1.8 million to oil-friendly federal politicians during the 2016 election cycle, with more than 90 percent going to the Republicans, who had dutifully shielded Exxon from carbon taxes and pollution regulations. During the six election cycles when he was CEO, nine of ten dollars donated by his company's PAC went to GOP candidates.

Exxon's corporate culture is not an admirable template for American idealism. Exxon already *is* a petro-state, wealthier than most countries, with its own private armies and intelligence apparatus. Now the head of Exxon is running our foreign affairs, with access to the many intelligence services and the capacity to bully states who don't tow the oil line.

Waterkeeper Alliance is a clean water advocacy group, of which I serve as president. Waterkeeper, which works in thirty-eight countries, has submitted a fifty-four-page petition to the EPA calling for the agency to enforce "bad corporate actor" rules and end all its federal contracts with ExxonMobil. The petition addresses Exxon's decades of deliberate lies—the company's campaign to deceive the public, politicians, and regulators about the danger of climate change. Recently-released documents prove that the sociopaths, including Tillerson, who ran Exxon knew for decades that its business activities would cause catastrophic climate change and mass death. Putting profits before people, Exxon kept its climate change science secret, while funding professional liars and nurturing the growth of a generation of climate change deniers. Under Rex Tillerson's leadership, the company continued to push government policies that buck proven science, human welfare, national security, and fundamental moral, ethical, and religious tenets. Last year, Exxon claimed as assets $330 billion in underground oil reserves that include some of the dirtiest fuels on Earth. The Securities and Exchange Commission and several states' attorneys general, led by New York's Eric Schneiderman, are currently investigating Exxon's failure to disclose to its stockholders the risks it has long known are posed to company value by the reality of global warming. According to Schneiderman, unless we are willing to write off planet Earth, about two-thirds of those reserves can never leave the ground. Exxon is therefore exaggerating its market value by hundreds of billions.

Tillerson has never expressed concern or even the slightest self-awareness that Exxon's business model threatens the future of humanity and life on Earth. America's largest oil company has accounted for more than 3 percent of global climate pollution, dating back to the mid-1800s. After years of putting Exxon's stock value ahead of humanity, will Tillerson now put America and the planet first? Tillerson's company would be severely impacted by the Paris Climate Accord to limit the burning of fossil fuels. His thoughts on climate change? "What good is it to save the planet if humanity [read Exxon] suffers."

And Tillerson didn't waste any time as head of the State Department to scrub the website of the Office of Global Change to reflect his stance. As noted by the Environmental Data and Governance Initiative, the revised website removed any mention of President Obama's Climate Action Plan to "reduce carbon pollution, promote clean sources of energy that create jobs, protect communities from the impacts of climate change and work with partners to lead international climate change efforts."

THE ENVIRONMENTAL PROTECTION AGENCY: SCOTT PRUITT

"These have power to shut heaven, that it rain not in the days of their prophecy: and have power over waters to turn them to blood, and to smite the earth with all plagues, as often as they will. . . . And men were scorched with great heat. . . . And every island fled away, and the mountains were not found."

—***Revelation,*** 11:6, 16:9,20

Trump's choice to run the EPA is an unctuous acolyte of Oklahoma's factory meat and Big Oil barons. Scott Pruitt built his career as a patsy for polluters: Prior to Pruitt's election in 2010, the Oklahoma attorney general's office had built a model environmental enforcement division under Kelly Hunter Foster, who is now a staff attorney for my organization, Waterkeeper Alliance. Foster had filed a dozen lawsuits against the poultry and industrial pork industries, which were polluting Oklahoma's air and waterways, sickening its citizens with effluvia of factory meat production, and putting family farmers out of business. Pruitt was the chicken industry's handpicked attorney general. Oklahoma's corporate meat barons financed Pruitt's campaign to rid themselves of Foster's lawsuits. Once in office, Pruitt dutifully terminated Hunter Foster's unit and shelved her docket. As attorney general, he never filed another environmental action. Instead, Pruitt turned his office's big guns against the EPA, filing a battery of federal lawsuits against the agency to challenge the Obama administration's anti-pollution and climate safeguards. These included suing the EPA to block the Clean Power Plan and another suit aimed at gutting rules on methane emissions from the oil-and-gas sector. "He let polluters off the hook and destroyed a decade of work," recalls Hunter Foster. "He has no environmental experience and no conservation instincts. His only qualification for his new job was his fierce hatred for EPA." Since his ascension to the administrator's post, Pruitt has frozen all new permits and scientific studies and put the agency in lockdown. He has promised to lay off 3,000 of the 15,000 EPA workers and cut the agency's already anemic budget by 31 percent, more than any other agency.

"And the merchants of the earth are waxed rich . . . for thy merchants were the great men of the earth; for by thy sorceries were all nations deceived."

—***Revelation,*** 18:3,23

Calvin Coolidge famously remarked that "the chief business of the American people is business." Trump has made it clear that business is to be the EPA's business as well. Pruitt burnished his resume for the EPA post with a major push by his mentor, Carl Icahn, a billionaire Wall Street hedge fund titan and generous Trump campaign donor. Icahn's holding company does business with the Koch brothers and TransCanada's Keystone pipeline system. A noisome EPA had accused Icahn's Oklahoma-based oil company of violating environmental laws. Based on these qualifications, Trump appointed Icahn to vet the contenders for the top-level EPA jobs.

Pruitt also received a boost from another of the Horsemen featured in this book—Oklahoma billionaire Harold Hamm (see Chapter 6). Hamm chaired Pruitt's 2013 reelection campaign. During the 2016 presidential election, Hamm had served as candidate Trump's energy advisor, but declined the president-elect's offer to head the Department of Energy.

Pruitt also boasts a direct Koch connection; as Oklahoma attorney general, Pruitt was simultaneously a director of the nonprofit Rule of Law Defense Fund, which received $175,000 in 2014 from a dark money umbrella group called Freedom Partners, the Koch network's political arm.

President Trump evidently shares Pruitt's antipathy toward the environmental agency. Upon announcing Pruitt's appointment, Trump added, "For too long, the Environmental Protection Agency has spent taxpayer dollars on an out-of-control anti-energy agenda that has destroyed millions of jobs." In mid-March, the president announced that he'd ordered Pruitt to revise one of President Obama's primary climate change policies—the EPA's strict standards on tailpipe pollution from motor vehicles. "As to climate change," Trump's director of the Office of Management and Budget said at a White House briefing, "I think the president was fairly straightforward; 'We're not spending money on that anymore.'"

On March 2, Pruitt told CNBC News with his characteristic "dumb as I wanna be" glee that humans were not responsible for global warming. Pruitt was proudly jockeying the EPA into position as the flagship of the new administration's anti-science crusade. The Bush administration had regarded science as a vanity of the despised liberal elite. One anonymous White House official, speaking to investigative journalist Ron Suskind, famously disparaged the liberal obsession with science-based inconvenient truths like climate change as "fact-based reality." But the Trump clown team has immediately achieved a new dimension of unhinged, by appointing a science-hating flat-earther as head of the world's premier environmental agency.

Even Christie Todd Whitman, who presided over the gutting of the EPA under George W. Bush from 2001 to 2003, was sickened by Pruitt's appointment. "I don't recall ever having seen an appointment of someone who is so disdainful of the agency and the science behind what the agency does."

Pruitt will have help from above as he plows under the rubble of his despised agency. In late December, Trump named Carl Icahn to a new administration position created by the president: Special Adviser on Regulatory Reform. While the administration proceeded to freeze adopting other new regulations, Icahn quickly succeeded in obtaining a special IRS rule that gives a tax break to his oil-refining company, CVR

Energy. Icahn is simultaneously pushing for a regulatory fix that would revamp an EPA rule (the Renewable Fuel Standard), which currently makes refiners responsible for ensuring corn-based ethanol is properly mixed into gasoline. Eliminating that requirement would have saved his company more than $200 million last year. Icahn, whose $16.6 billion is a fortune larger than all the other cabinet members combined, claims immunity from such conflict-of-interest problems because he's simply an "unpaid adviser" to the administration.

SECRETARY OF INTERIOR: RYAN ZINKE

"And there followed hail and fire mingled with blood, and they were cast upon the earth: and the third part of trees was burnt up, and all green grass was burnt up. . . . And the third part of the creatures which were in the sea, and had life, died. . . . And the sun and the air were darkened by reason of the smoke of the pit."

—*Revelation,* 8:7,8, 9:2

My friend, Leonardo DiCaprio, a leading climate activist, gave a presentation to Trump soon after the election. He and DiCaprio Foundation president Terry Tamminen, the former Santa Monica BayKeeper and chief of California EPA under Governor Arnold Schwarzenegger, unveiled a plan for creating millions of jobs by encouraging the growth of clean, renewable energy. Looking at the plan approvingly, president-elect Trump told Leo that he wanted to be the twenty-first-century Teddy Roosevelt. Leo gave him a copy of his new documentary *Before the Flood* describing the perils of climate change, and the president-elect promised to watch it. Afterward, Leo learned that Trump's team had announced the appointment of Scott Pruitt, while they were still in the meeting. Trump had warned Leo, "There are going to be some you will consider bad appointments." But, he promised the actor, "You're really gonna like who we put in for Interior."

That environmental superhero turned out to be Ryan Zinke, a first-term congressman from Montana who also describes himself as "a Teddy

Roosevelt guy." But while Roosevelt dismantled Standard Oil, Zinke has spent his career suckling at the industry teat, gagging down $345,136 of oily money from petro interests. In the House, Zinke represented the Powder River Basin, a once edenic wilderness, transformed into a moonscape by federal coal-leasing policies, championed by Zinke. In fact, in recognition of his enthusiasm as a cheerleader for coal extraction, the League of Conservation Voters awarded Zinke a 3 percent score. In 2008, Zinke said he believed in climate change, but has since dutifully recanted, in goose-step with the Republican Party leadership. It isn't "proven science," he now insists.

SECRETARY OF ENERGY: RICK PERRY

"And lo, there was a great earthquake; and the sun became black as sackcloth of hair, and the moon became as blood."
—*Revelation*, 6:12

Modest support and research progress for wind and solar efficiency have long made the Department of Energy a bugaboo to the fossil fuel cartel. Four years ago, Rick Perry, the former Texas governor, promised to abolish the department that President Trump has now appointed him to lead. Oil and gas tycoons funded his two presidential campaigns. The CEO of Texas-based Energy Transfer Partners (ETP), Kelcy Warren, the owners of Dakota Access Pipeline, donated $5 million to a pro-Perry super-PAC during the 2016 race. When Perry's run for the White House fizzled, he accepted a sinecure as a paid board member of ETP (receiving $236,820 in 2015). Warren went on to contribute $103,000 to Trump's campaign. Perry sold his shares after the election to avoid "conflict of interest." President Trump also had a personal stake in that notorious pipeline. He invested nearly $1 million in Energy Transfer Partners in 2015, and between $250,000 and $500,000 last year in Phillips 66, which owns a 25 percent stake in the catastrophic boondoggle. Perry quickly proved himself a trustworthy manager of Trump's investment by presiding over the project's resurrection.

CIA DIRECTOR: MIKE POMPEO

"Four beasts full of eyes before and behind."

—***Revelation,*** 4:6

Kansas Congressman Mike Pompeo, an errand boy appointed by the Koch brothers to represent their Wichita hometown in Congress, is President Trump's CIA chief. He received more campaign donations from the Kochs than any other member of Congress. Pompeo is a Tea Party shill, so enthralled by the Warfare State and so out of touch with American values that from his seat on the House Intelligence Committee, he paid sunny homage to the CIA's brutal detention and interrogation programs. Pompeo's chief of staff, Jim Richardson, was a former Koch lobbyist. When Pompeo entered the House in 2011, his first order of business was opposing the Obama administration's plan to create a public EPA registry of greenhouse gas polluters. Later, Pompeo introduced legislation to kill tax credits for wind power, saying it should "compete on its own," an idea aggressively promoted by the Koch brothers, whose oil and coal enterprises are heavily subsidized by billions of dollars in federal taxpayer lucre. (A recent report by the International Monetary Fund calculates global energy subsidies at over $5 trillion annually, with the United States providing $700 billion in subsidies to Big Oil, the richest industry in the history of the planet.)

Foreign policy experts warn that close historical ties between the CIA and the oil industry have led America into its most catastrophic foreign policy disasters. Instead of serving the American people and our national historic ideals, the agency has, since the days of Allen Dulles, a former oil company lawyer, routinely deployed its awesome power to serve the mercantile interests of oil companies and US-based multinationals. Our volatile relationship with Iran began in 1953, when the CIA overthrew Mohammed Mossadegh, that region's first democratically elected leader in four thousand years, as a favor to US and British oil corporations, which derailed Mossadegh's plan to allow his people to benefit from Iran's oil resources. Catastrophic blowback from that dark episode continues to reverberate across the Middle East today. The Iraq War—fueled in part by US eagerness to grab control of Saddam Hussein's oil fields—not only killed nearly 4,500 Americans, but continues to destabilize the entire region. The Syrian civil war and refugee disaster are largely the result of the CIA taking sides in a pipeline dispute between Sunni and Shia, a debacle that fueled the creation of ISIS. The unseemly ties between Big Oil and the intelligence service of this country have stained our reputation, eroded our moral authority, made us the target of terrorist attacks, and led us into costly blowbacks, with surcharges in American blood, treasure, and prestige beyond calculation. At a time in our history when we should be de-coupling our foreign policy from Big Oil, we've got the oil industry's most visible tycoon since John D. Rockefeller *running* the State Department and an oil man's sock puppet *running* the CIA.

> *"And it was commanded them that they should not hurt the grass of the earth, neither any green thing, neither any tree."*
> —***Revelation***, 9:4

During his first weeks in power, President Trump kept his promises to the carbon cartel. I watched thirty-three years of my work reduced to ruins as the president mounted his assault on science and environmental protection. The new administration hit the ground running, announcing plans to eliminate funding for NASA's climate research programs. At EPA, president-elect Trump's transition team launched a Soviet-style purge of climate change scientists, demanding a list of every employee or contractor who had attended meetings of the Interagency Working Group on the Social Cost of Carbon—and all materials that were distributed or generated afterward. Trump advisors demanded that Foggy Bottom diplomats disclose monies that the State Department provided to international environmental groups (for example, as part of the Paris Climate Agreement, the United States pledged $3 billion to a Green Climate Fund designed to help poor countries develop renewable energy and adapt to climate change)—presumably to target them for erasure.

In response to these attacks, a newly created Climate Science Legal Defense Fund published a guide for government researchers targeted for attack and censure as its government scientists worked feverishly to preserve decades of critical research.

The new president quickly signed legislation abolishing rules that forbade coal companies from filling streams and buffer zones with mining waste, putting thousands of miles of rivers and streams at risk from this formerly outlawed practice. He issued an executive order to rescind a 2015 rule aimed at protecting small streams and wetlands, and a directive to abolish protracted environmental reviews.

He announced measures to stop the environmental impact statement required before the Dakota Access Pipeline could be constructed, and then instructed the Army Corps of Engineers to give Energy Transfer Partners an easement to forge ahead with the tragic project across Sioux lands. Trump also set out to resurrect the Keystone XL oil pipeline stopped by the Obama Administration after a massive public outcry. In late March, Tillerson's State Department approved moving forward with construction. The 1,700-mile-long pipeline will propel an estimated 35 million gallons of dirty oil every single day from Alberta, Canada's Tar Sands, across American aquifers to refineries on the Gulf Coast. The Koch brothers hold close to two million acres of those tar sands, more than the combined assets in the area of ExxonMobil, Chevron, and Conoco.

The new president instructed the EPA to promulgate regulations to kill Obama's hard-won Clean Power Plan, the law that finally restricted carbon emissions from power plants. The new administration moved to open protected federal lands for drilling and mining, and to lift a moratorium on coal leases on federal lands.

In mid-March, Trump directed the EPA to get rid of another of Obama's signature achievements—stringent fuel economy standards passed to help meet America's international commitment to cut carbon emissions. This drastic policy reversal not only makes it impossible for the United States to comply with the Paris Accord but also jeopardizes America's booming lead in the electric vehicles industry.

The next day, Trump sent a proposed budget to Congress that would slash the EPA's funding by 31 percent and lay off about one-fifth of its staff. The White House will cut the climate protection budget by nearly 70 percent to $29 million. Virtually eliminated are the environmental justice program, established in 1992, and the Chesapeake Bay program, established in 1983 to clean up the largest estuary in North America. Also on the scrap heap is the Energy Star program that has saved consumers an estimated $430 billion on their utility bills and avoided 2.7 billion metric tons of greenhouse gas emissions. President Trump's so-called apocalypse budget guts funding to the United Nations for its climate change efforts, cuts 17 percent of the National Oceanic and Atmospheric Administration's climate data program, and eliminates the Sea Grant program that prepares coastal communities for storms and sea level rise. The proposed State Department budget contains zero funding for international climate negotiations. While slashing America's environmental protection, Trump moved to pump up military spending by another $54 billion.

Meantime, Scott Pruitt got busy distributing his top EPA jobs to America's most vehement climate science skeptics, including three former staffers of Oklahoma Senator James Inhofe, who called global warming "the greatest hoax ever perpetrated on the American people." In his first speech to EPA employees, Pruitt scolded his new employees to improve the agency's relationship with private businesses. He omitted all mention of protecting public health or the environment. In a speech before a gathering of conservatives, Pruitt applauded the polluters who want to eliminate his agency altogether as "justified": he added, "People across the country look at the EPA the way they look at the IRS."

Pruitt, interviewed on CNBC on March 9, reassured the network that carbon emissions are not, after all, the "primary contributor" to climate change. That day, NOAA announced that the levels of CO_2 in the atmosphere rose at a record pace for the second year in a row, and President Trump announced his scheme to zero-out NOAA's climate change research budget.

Trump and Pruitt have expressed intense hostility toward federal protection of the environment. They mean to return us to the era before federal environmental laws; the era before Earth Day when states were engaged in a wholesale race to the bottom to eliminate regulation to recruit filthy industries. The era when rivers caught fire and pollution killed tens of thousands of animals annually, DDT exterminated entire populations of birds, and Lake Erie was declared dead. With no federal control, states will once again compete with each other to become pollution havens in exchange for a few years of pollution-based prosperity. This hijack of American democracy by oil tycoons is a suicide pact for our planet. We are already living in a science fiction nightmare when all credible scientists are saying that their former predictions on global warming were radically conservative. The cataclysms they warned would happen in a century or two are happening now. The Intergovernmental Panel on Climate Change cautions that if we conduct business as usual—our current course—our planet will experience a six-degree Celsius temperature rise by the turn of the next century. The last time the Earth was six degrees warmer, crocodiles lived at the North Pole.

Today, there are more than seven billion humans on the planet, few of whom will be able to adjust and survive the concomitant floods, storms, typhoons, and hurricanes intensified by climate change. Children currently alive will suffer dystopian global upheavals beyond human experience, or the capacity of organized civilizations to endure. Our great coastal cities will be drowned by sea level rise; multitudes will starve when lands become arid and lifeless; homes and businesses will succumb to forest fires of increasing frequency; children will suffer or die from the side effects of insect-borne diseases, such as microcephaly, spreading rapidly to formerly temperate regions of the Earth as tropical insects multiply; millions will suffer from food shortages as crops fail due to changing climate conditions. These impacts emerge straight from The Book of Revelation, extreme weather on a biblical scale—destructive droughts, lethal superstorms, floods, fires, melting glaciers, rising seas, drowning cities, and disappearing species.

Whether we recognize it or not, we are all locked in a life and death struggle with these corporations over control of both our landscapes and political sovereignty. The Kochs' corporate vision for our country would commodify not just the land, the air, and the water but also our people. Everything we value becomes expendable in their drive for corporate profits. "To greed," Seneca observed, "all nature is insufficient." That hunger will devour our people, our natural world, and the other assets of our patrimony. Corporate efforts to privatize the commons are occurring in all parts of the world, and it's no accident that environmental injury correlates almost perfectly with political tyranny; and those carbon tyrants would steal from us our air, our water, our wildlife, fisheries, and public lands, the shared resources of our society—the commonwealth assets that provide the *gravitas* around which communities coalesce.

The battle against Trump and his Horsemen is not just a battle to protect our waterways, our livelihoods, our property, and our backyards. It's a struggle for our sovereignty, our values, our health, and our lives. It's a battle for dignified, humane, and wholesome communities. It's a defensive war against toxic and economic aggression by Big Oil and King Coal. It's a struggle to break free from the merciless tyranny of the carbon cartel and create an economic and energy system that is fair and rooted in justice, economic independence, and freedom.

If we're to leave behind a habitable world, the Horsemen need to be reined in, bridled, and broken.

Robert F. Kennedy, Jr. is President of the Waterkeeper Alliance.

Chapter 1

The Choices Before Us

Early in October 2015, Rex Wayne Tillerson—62-year-old father of four, recent national president of the Boy Scouts of America—took the stage at the 36th annual Oil and Money Conference in London. As the then-CEO of ExxonMobil, Tillerson had just been named Petroleum Executive of the Year. His topic was "Unleashing Innovation to Meet Our Energy and Environmental Needs."

Tillerson's half-hour-long speech did not ignore the subject of a rapidly changing global climate. He spoke of the challenge of "reducing the greenhouse gas emissions associated with energy use." He said "the risks of climate change are serious and warrant thoughtful action," including his corporation's research into alternative technologies and support of a "revenue-neutral" carbon tax. However, Tillerson added, "The world will need to pursue all energy sources, wherever they are economically competitive . . . importantly, we will need coal, oil, and natural gas."

The highest paid executive of the richest fossil fuel corporation on the planet went on to point out: "From the very beginning of concern on this

71

issue, ExxonMobil scientists and engineers have been involved in discussions and analysis of climate change. These efforts started internally as early as the 1970s."

What Tillerson failed to mention was this: only the month before, an investigation of internal Exxon documents had revealed that those very scientists had repeatedly warned, almost forty years ago, of a potentially "catastrophic" warming of the planet that "endangered humanity." But instead of responding to this red alert from their own experts by starting to shift the energy giant toward renewable resources, Exxon's top executives, including Tillerson, had shut down the company's own research—and embarked instead on a massive disinformation campaign aimed at debunking climate change as a myth.

The corporation was a ringleader in setting up the Global Climate Coalition, a massive disinformation machine bringing together the world's leading fossil fuel companies in an all-out effort to prevent governments from curbing their emissions. Tillerson's company, the second largest emitter of CO_2 in the world (after Chevron), dispersed millions to muddy any scientific understanding and delay any real action.

Tillerson, his predecessor Lee Raymond, and their cronies knew the truth about the fate of the planet. And yet they lied, and they paid others to lie. They lied as global temperatures began rising at record rates. They lied as droughts and wildfires swept across the American West, and as California started running out of water. They lied as tornadoes and hurricanes and snowfall levels intensified in unprecedented ways. They lied as thousands died in European heat waves, and thousands more perished in Asian floods. They lied as Greenland's ice turned liquid, and sea levels began to rise two-and-a-half times faster than anyone thought possible, and the oceans became increasingly acidic and filled with disease-causing bacteria. They lied and sacrificed future generations for their short-term profits.

During his visit to America in December 2015, Pope Francis issued a warning about climate change, "a problem which can no longer be left to a future generation. . . . I can say to you 'now or never.' Every year the problems are getting worse. We are at the limits. If I may use a strong word I would say that we are at the limits of suicide."

Six months earlier, in the pope's encyclical on the situation, he had asked: "What kind of world do we want to leave to those who come after us, to children who are now growing up?" And he had raised another question: "What would induce anyone, at this stage, to hold on to power only to be remembered for their inability to take action when it was urgent and necessary to do so?"

The president of the World Bank, Jim Kim, has spoken out along similar lines: "My son will live through a 2, 3 or maybe even 4 degree Celsius warming. We cannot keep apologizing to our children for our lack of action. We must change course now."

At the 2016 Summer Olympics in Rio de Janeiro, while some three billion people watched, the opening ceremony featured a video of ever-escalating global carbon pollution and simultaneously drastic rise in sea levels.

These are the facts behind the pleas of the Pope, the World Bank leader, and the Olympic Games leadership:

- Sixteen of the seventeen warmest years ever recorded have occurred since 2001. For the third consecutive year, it was announced in January, the earth set a heat record. Across vast stretches of the Arctic Ocean, temperatures in the fall of 2016 reached an astonishing 20 to 30 degrees above normal.
- As billions of tons of ice melt or slide into the sea, satellite data shows that oceans around the world are rising by five millimeters a year, a rate not seen since the close of the last Ice Age.
- "Much of the carbon we are putting in the air from fossil fuels will stay there for thousands of years—and some of it will be there for more than 100,000 years."—Oregon State University paleoclimatologist Peter Clark, lead author of a new study in *Nature Climate Change*, February 2016.

- "Given currently available records, the present anthropogenic carbon release rate is unprecedented during the past 66 million years." —*Nature Geoscience*, March 2016.

Several years earlier, in September 2013, UNICEF published the results of a five-year study about how a changing global climate affects today's children. "Climate change has too often been discussed and debated in abstract terms, negating the human costs and placing little attention on its intergenerational impact," the report said. However, "more severe and more frequent natural disasters, food crises and changing rainfall patterns are all threatening children's lives and their basic rights to education, health, clean water, and the right food."

These drastic changes in our planet's ecosystem will have the most severe consequences, of course, on future generations. Climate change is all too often discussed in "abstract terms," the 2013 UNICEF report noted. But the environmental upheaval associated with climate change is already having a massive impact on "children's lives and their basic rights to education, health, and [proper] food." UNICEF has estimated that, by 2030, 25 million more children will suffer malnourishment, with another 100 million facing food insecurity due to scarcity, and between 150 and 200 million more being displaced from their homes. "We are hurtling towards a future where the gains being made for the world's children are threatened, and their health, wellbeing, livelihoods and survival are compromised . . . despite being the least responsible for the causes," said David Bull, executive director of UNICEF in the United Kingdom.

The UNICEF report noted that "children and young people in developed countries are acutely aware of climate change, and are passionate and vocal about the need for action by governments to tackle the problem." Polling in the UK indicated that nearly three-quarters of those between ages 11 and 16 in Britain worried about the planet's environmental future. More than seven in ten wanted their government to do more, and nearly two-thirds voiced particular concern about their counterparts in developing nations. In the US, similar polling found almost three-quarters of young voters saying they were less likely to vote for a candidate who opposed President Obama's climate change plan. "We need to listen to what children are saying," the study concluded.

The goal of the entrenched interests, however, is to drown out those voices—all the way to the classroom. In Wyoming, when the Park County School District was to vote on whether to purchase new textbooks and reading materials in 2015, one board member responded, "I will *not* authorize any of the $300,000 allocated for this purchase to include supplemental booklets about 'global whining'. . . . Our Wyoming schools are largely funded by coal, oil, natural gas, mining, ranching, etc. This junk science is against community and state standards."

Jeff Turrentine, who wrote about this for *OnEarth Magazine*'s web site, added, "For thousands of years, going back to Aristotle, humanity's greatest minds have sought to safeguard the precepts of the scientific method by keeping them away from the corrupting influence of political culture. Defending the integrity of science from powerful people is what got Galileo imprisoned. And yet, 400 years later, here we are: watching a public official tasked with guiding the educational trajectories of his community's children rail against the accepted science on climate change—because its conclusions threaten to undermine the local political culture. . . . Anyone who would deliberately misinform children about the gravity of the problem that awaits them when they grow up doesn't deserve to be in charge of their education."

The campaign to "misinform children" is particularly aggressive in the American West, stronghold of the oil and coal industries, including in Utah, where a coalition of parents decrying "Education Without Representation" has intimidated the state's Office of Education into watering down education on climate change. Even in "left-coast" California, where the Democratic Party has a lock on state government, a 2015 analysis of science textbooks used in the sixth-grade classrooms revealed that the language and writing techniques "more closely match the public discourse of doubt about climate change rather than the scientific discourse." The study, which was conducted by Southern Methodist University, speculated that conservative media like Fox News had contributed to "a shift in public discourse, which eventually influences textbook language by creating competing interests within the textbook market." A follow-up survey published in the journal *Science* in 2016 found that, while three-quarters of science teachers nationwide devote time to climate change instruction, 30 percent tell students that it's "likely due to natural causes" and another 31 percent claim that the matter is unsettled. That's opposed to the 97 percent of active climate scientists who contend that human activity is a primary cause. Bills have now been introduced in state legislatures of four states that promote climate change denial as part of academic freedom.

Even in 2016, as the world weathered another year of record-setting temperature rise, America's presidential campaign was dominated by Republican candidate Donald Trump, who dismissed the global crisis as a "hoax"—allegedly manufactured by the Chinese in order to make US manufacturing "non-competitive" and by Democrats to justify higher taxes. Meanwhile, Trump's Democratic opponent, Hillary Clinton, while acknowledging climate change as an urgent problem, amassed a huge campaign war chest from donors and lobbyists connected to the oil, gas and coal industry, while her allies headed off an attempt by Senator Bernie Sanders to hammer an anti-fracking plank into the 2016 party platform.

Despite all the calamitous news from the environmental front lines, the energy industry still wields extensive influence over the climate change debate, from the classroom to the presidential campaign trail.

How does this terrifying disconnect happen? How do our political and corporate leaders continue to defy scientific reality and mislead the public, even though the consequences of this failed leadership will certainly be disastrous for future life on the planet? How do men like Rex Tillerson explain themselves to their own children and grandchildren, inheritors of the epic environmental havoc being brought about by him and other energy moguls? As carbon dioxide has risen to atmospheric levels not witnessed on earth in millions of years, a relative handful of men have fought to maintain their power and wealth at the expense of all civilization. This book scrutinizes who these people are, their means of confusing the truth, and how they justify their actions.

You will learn about the long, sordid history of ExxonMobil's cover-up of the looming climate-change disaster. You will learn about the clever propagandists hired by the energy industry—men like Richard Berman, the founder of a Washington-based PR firm that launches nonprofit "charitable" front groups aimed at derailing any government efforts to put the brakes on the worst carbon polluters. "Factual debates," Berman once said in a speech, often leave people "overwhelmed by the science." But if you could get enough people on your side, you could create "a position of paralysis about the issue. . . . They don't know who is right." Berman's son David, a singer-songwriter in Nashville, knows who is right. David Berman has publicly proclaimed that his father is "a sort of human monster"—and he himself is the "son of a demon come to make good the damage." And so you will also hear from the children and grandchildren of these destroyers of life, some of whom struggle with an extra burden of familial responsibility.

You'll learn of the damage being done in Oklahoma, where the release of the most potent greenhouse gas, methane, is accelerating the climate problem—and the underground disposal of water associated with oil and gas production is causing earthquakes at rates never seen before. The tracks of this attempted cover-up lead straight to Harold Hamm, billionaire CEO of America's largest energy independent, Continental Resources. Hamm donates millions to the University of Oklahoma, which happens to host the state's Geological Survey. When its seismologists sought to sound the alarm about fracking's connection to the rapid increase in quakes, Hamm set out to try to get them fired.

You'll learn about the hypocritical "Energy Poverty" campaign put forth by Peabody Energy's CEO Greg Boyce, as justification for vastly expanding its coal production into impoverished countries around the world. You'll see how the coal barons have heaped praise (and poured dollars) into the bank accounts of US Representative Lamar Smith of Texas for spearheading an investigation into scientists and environmental officials who assert that climate change is a reality.

You'll be taken inside the domain of the infamous Koch brothers, whose underhanded campaign in their native Kansas has stymied efforts to use wind energy as a primary power source. Charles and David Koch have spent a fortune buying off political candidates nationwide and funding the American Legislative Executive Council (ALEC), whose State Policy Network pushes bills aimed at slowing down deployment of renewable energy sources.

Energy titans like the Kochs have no hesitation when it comes to protecting their enormous wealth, even if it means going after supreme religious leaders. Before Pope Francis came to Philadelphia in 2015, the Heartland Institute —a nonprofit funded by the Koch brothers, ExxonMobil, and other big corporations—called a press conference to denounce the pontiff's impassioned climate change encyclical. "What is environmentalism but nature worship?" Heartland's marketing director asked. "I'm wondering as a scholar if pagan forms are returning to church." It was a wild and absurd punch, thrown at one of the most beloved public figures in the world. But the dark lords of fossil fuels have profited extravagantly from muddying their opponents and the climate change discourse.

Free speech, if these people have their way, is an endangered species. Florida's Republican Governor, Rick Scott, another faithful servant of the energy industry, has ordered his Department of Environmental Protection not even to use the terms "climate change" or "global warming" in any of their official communications, emails, or reports. This is a state where scientists are warning that sea-level rise could inundate much of the coastline over the next century, including great swaths of metropolitan Miami. In Wisconsin, where Governor Scott Walker— another favorite of the Koch brothers—runs the show, the Board of

Commissioners of Public Lands has gone even further. Its staff was banned in 2015 from even *discussing* climate change.

During the historic December 2015 climate conference in Paris, representatives of 195 countries came together to reach a landmark accord. For the first time, there was a sense of real urgency among world leaders about what the burning of fossil fuels is doing to our planet, a political shift of global proportion that's been a long while coming even if it doesn't go far enough. But the climate deniers were also on-hand in Paris, spreading their propaganda at a "Day of Examining the Data" counter symposium held in a hotel across town from where the United Nations negotiations took place. It was co-sponsored by the Heartland Institute and the Committee for a Constructive Tomorrow (CFACT), which is also funded by the fossil fuel companies.

Chris Horner was there, a lawyer who's filed numerous suits harassing legitimate climate scientists, operating behind-the-scenes on behalf of the coal companies. So was Marc Morano, once Director of Communications for US Senator and climate-change-denier James Inhofe and now running the anti-science Climate Depot website. Morano was busy promoting CFACT's propaganda documentary film, "Climate Hustle," aimed at confusing the public about whether the climate science is real. A Heartland spokesman described the event as "successful." In truth, it turned out to be an embarrassing flop, with only about 30 people in attendance (about twice the number of speakers).

If the dark alliance between corporate and political power makes climate change progress often feel hopeless, there are nonetheless rays of light. As musician-activist David Berman has demonstrated, the war to save the planet has literally come home, with a growing number of "next-gen" members of fossil fuel families throwing themselves into the struggle for climate-change action, even if it brings them into direct conflict with their fathers and grandfathers and the energy companies they started or manage. A decade ago, a dozen descendants of John D. Rockefeller—founder of the oil empire that eventually became Exxon—began pressuring the company to cease funding climate change deniers and to invest in renewable energy. The Rockefeller Foundation has gone so far as to divest its stock in ExxonMobil. And shareholder pressure has prompted Tillerson's company to pull away from funding several of the leading purveyors of doubt.

Christopher Lindstrom, a great-great-grandson of the Standard Oil founder, has gone a step further. He's pouring his inheritance into what he calls regenerative bioenergy. Lindstrom attends the annual conference of a new organization called Nexus, held at the United Nations and bringing together some 500 millennials from around the world seeking new ways and means to invest their inherited wealth. *Horsemen of the Apocalypse* examines the alternative solutions that some of the energy renegades are working on. Among others, you'll meet Katherine Lorenz, whose grandfather made his fortune pioneering natural gas fracking, but whose philanthropic foundation is dedicated to moving Texas toward a clean energy portfolio.

It's time to hold the perpetrators accountable, end their reign of lies, and bring about the massive change in global energy policy that they have long been resisting. As *Horsemen of the Apocalypse* reveals, this crusade to save civilization pits the energy moguls against a younger generation that is using the courts and civil disobedience while seeking alternative sources of fuel and electricity. And the outcome of this intimate struggle will determine the fates of many generations to come.

Chapter 2

Rex Tillerson, Lee Raymond, and the Duplicity of ExxonMobil

About a 45-minute drive north of Dallas, bordered by the communities of Flower Mound and Double Oak, lies the small Texas town of Bartonville (population: 1,469 as of the 2010 census). Besides the main drag, oddly still called 407 Farm Road, the designation "rural" applies mainly to Bartonville's flaunted association with thoroughbred horses. Saddlebrook Estates, guarded by a pair of equestrian statues at its non-gated entry, hosts a "premier boarding and training facility" at the beginning of its subdivision of sprawling, custom-built homes. From Kentucky Derby Drive, you soon reach Triple Crown Court and Noble Champions Way.

Whether former ExxonMobil CEO and Chairman Rex Tillerson lives somewhere in Saddlebrook Estates is not easy to determine. Even locating the Bartonville Town Hall requires some diligence. "I always forget about it because it's so small," says a woman who works at the Lantana Trail development across the highway. But there it is, just past the closed Exceleron gas station and the also-shuttered Bartonville Food Store (supplanted by the nearby, upscale Town Center shopping mall a decade or so ago).

"Our median home value is about $460," the lone clerk on duty at the Town Hall reveals, meaning in the thousands. She isn't sure whether the energy mogul whom Trump plucked to run the State Department might own a home in Saddlebrook. But she knows that his 83-acre Bar RR Ranch is less than a mile away at the edge of Farm Road. "You'll see it, three gray posts on the left," the clerk says. "There's a little pull-off for the driveway, but you can't turn in there, uh-uh. Just be careful at the ranch, he's got some security people there. I'm sure they're everywhere, but you never really kinda know."

Behind a locked wrought-iron gate, beyond the no trespassing signs and the "guard dog on premises" warning, the $5-million ranch property seems to stretch endlessly past the horse stables and training facilities. Atop the central post in the entryway, mounted in paving stone, is something else. It's a large sculpted metal globe, which looks like someone took a carving knife to it. The upper portion, along uneven serrated edges, opens into a gaping void.

It's Rex Tillerson's globe . . . a world whose apex has been systematically shredded into pieces. And now he is our Secretary of State.

Within minutes of the intrusion apparently having been observed, a white pickup truck rolls up at the gate, its driver making clear that prying eyes are not welcome.

Rex Wayne Tillerson, a product of small towns in Texas and Oklahoma whose father worked four decades as a professional organizer for the Boy Scouts of America, is himself an Eagle Scout. Indeed, Tillerson still denotes this on his resumé. His corporate speeches often cite the Scout Oath and Scout Law. After assuming leadership of the world's largest private corporation in 2003, Tillerson initiated a program modeled upon the Boy Scouts' merit badge system, in which employees receive medals or coins for exhibiting teamwork and leadership skills. But when it comes to his business practices, Tillerson is not a good Scout. Doesn't a Boy Scout leave a place cleaner than they found it?

Tillerson has shown he can be a good steward of the earth, however, if he happens to own the plot of earth. He leaped into action in 2014, when a 15-story water tower was slated to be built adjacent to his ranch. The tower's purpose? Providing water to a nearby natural gas drilling site that utilized hydraulic fracturing, commonly known as "fracking."

ExxonMobil had lately become America's largest homegrown producer of natural gas, and Tillerson had publicly said of attempts to curtail fracking, "This type of dysfunctional regulation is holding back the American economic recovery, growth and global competitiveness." However, in his backyard, fracking would apparently create "a constant and unbearable nuisance to those that live next to it," including "traffic with heavy trucks" which would devalue his property. So Tillerson joined a lawsuit aimed at shutting down the project, along with co-plaintiff Dick Armey, the former Republican House Majority Leader. Adding to the richness of the irony, Armey had gone on to chair Freedom Works, a Tea Party group that loudly supports fracking. But Armey's $2 million, 78-acre ranch also lies adjacent to the water tower site, which puts things in a different perspective. In March 2014, a judge dismissed the claims brought by Tillerson and his wife Renda, but Armey and other plaintiffs hung tight in a suit against the Bartonville Water Supply Corporation.

Perhaps he wants to keep his ranch unspoiled for his children and grandchildren. But when it comes to the health of the planet, Tillerson has little concern for future generations. In August 2015, 21 young people from around the US filed a still-ongoing lawsuit alleging that the federal government has violated their rights by failing to protect present and future generations from human-caused climate change. Ranging in age from 8 to 19, the youth addressed conditions near their own homes: extreme drought, a threatened forest leading to water scarcity, an unswimmable river due to fish die-offs. ExxonMobil, Koch Industries, and dozens more oil and gas companies were worried enough about the children's crusade to join the government's effort to defeat the lawsuit. The fossil fuel powerhouses called the youth's case "extraordinary" and "a direct threat to [their] businesses." In response to the energy industry's aggressive counterattack, retired NASA scientist James Hansen, who has been sounding the alarm about climate change for decades, declared, "I am not surprised that fossil fuel corporations seek to derail this case, but the fundamental rights of my granddaughter and future generations to life, liberty, and pursuit of happiness must prevail."

Rex Tillerson joined Exxon in 1975, upon graduating from the University of Texas. He would rise quickly through the ranks of the multinational energy giant's oil-and-gas discovery division. Kenneth Cohen entered the corporation as legal counsel in 1977, ultimately going on to run its public affairs department. Cohen would also serve as a national trustee of the Boys and Girls Clubs of America. Tillerson and Cohen, so-called youth advocates, would become primary players in denying *to this day* the catastrophic risks that climate change poses to the very survival of future generations.

Nor are they alone. Jack Gerard has been at the helm since 2008 of the American Petroleum Institute, the industry's leading lobby. Gerard, who has been named one of Washington, D.C.'s "Power 100," and his wife have eight children, including twin boys adopted from Guatemala. He is a past chairman of the National Capital Area Council of the Boy Scouts of America, which hosts thousands of youth in Washington, Maryland, Virginia and the Virgin Islands, and he continues to serve as a Boy Scouts board member. But, like Tillerson and Cohen, there is nothing "morally straight"—the ethical commitment made by those who take the Boy Scout pledge—about Gerard when it comes to his dishonest declarations about the stark choices facing the human race.

In delivering a State of American Energy address in January 2016, Gerard spoke of how fossil fuels must "remain the foundation upon which our modern society rests for decades to come" despite "an ardent few who continue to believe that keeping our nation's abundant energy resources in the ground is a credible and viable national energy strategy. There are some in government who will advance their favored forms of energy to that dubious and untested end, heedless of the potential harm it could cause to our economy." Against all scientific wisdom, Gerard also demanded the elimination of all government obstacles to future carbon energy exploitation, decrying the "dangerous combination of outdated policies and anti-fossil fuel political ideology that discourages American companies from investing in tomorrow's pipelines, marine terminals and other energy infrastructure projects." Be prepared, Boy Scouts are taught. This has become a particularly urgent lesson as the human race is forced to prepare for one natural disaster after the next, linked to our changing climate.

All the way back in 1958, Bell Laboratories funded a series of TV science specials produced by the legendary Frank Capra. One of the episodes, titled "The Unchained Goddess," featured Dr. Frank B. Baxter, a professor at the University of Southern California. Over a half century ago, Baxter told a national audience, "Even now, man may be unwittingly changing the world's climate through the waste products of his civilization. Due to our release, through factories and automobiles every year, of more than six billion tons of carbon dioxide . . . our atmosphere seems to be getting warmer. It's been calculated that a few degrees rise in the earth's temperature would melt the polar ice caps, and if this happens, an inland sea would fill a good portion of the Mississippi Valley. Tourists in glass-bottomed boats would be viewing the drowned towers of Miami through 150 feet of tropical water."

Four years later, Humble Oil and Refining—to be "rebranded" in the early 1970s as Exxon—took out a two-page color ad in *Life Magazine*. Below a beautiful color photograph of Alaska's cloud-bedecked Taku Glacier, its headline read: "EACH DAY HUMBLE SUPPLIES ENOUGH ENERGY TO MELT 7 MILLION TONS OF GLACIER!" The small print continued: "This giant glacier has remained unmelted for centuries. Yet, the petroleum energy Humble supplies—if converted into heat—could melt it at the rate of 80 tons each second!"

Was it prescience? Or a kind of Freudian slip? We'll never know.

We do know from recently uncovered documents that, as early as 1968, the American Petroleum Institute received a report from the

Stanford Research Institute concerning "sources, abundance, and fate of atmospheric pollutants." It concluded that carbon dioxide emissions from fossil fuels were "outstripping the natural CO_2 removal processes that keep the atmosphere in equilibrium" and that "significant temperature increase could lead to melting ice caps, rising seas, and potentially serious environmental damage worldwide."

Industry scientists confirmed that urgent research was required to bring these emissions under control. Clifford Garvin, while CEO at Exxon between 1975 and 1986, decided to install solar panels for heating his swimming pool in the New Jersey suburbs. At the time, President Jimmy Carter did the same on the roof of the White House, while initiating a program aimed at the country getting 20 percent of its energy from renewable sources by the year 2000. Both these moves coincided with a 1979 National Academy of Sciences (NAS) study concluding that if manmade carbon dioxide emissions continued to grow, there was "no reason to doubt that climate changes will result and no reason to believe that these changes will be negligible . . . a wait-and-see policy may mean waiting until it is too late."

But wait-and-see quickly became the order of the day. With the election of Ronald Reagan in 1980, down came the solar panels from the White House. Within Exxon, the CEO's solar gesture was soon regarded as a prime example of what *not* to do; alternatives to petroleum simply weren't economically sustainable.

Even before the NAS study appeared, Exxon was well aware that something potentially disastrous was underway. This is one of the biggest corporate scandals in our country's history. As author and climate activist Bill McKibben, founder of 350.org, has said, "Even as someone who has spent his life engaged in the bottomless pit of greed that is global warming, the news and its meaning came as a shock: We could have avoided, it turns out, the last quarter century of pointless climate debate."

Here's where the obfuscation began: At a July 1977 meeting inside the company's then-headquarters in New York City, Exxon senior scientist

James F. Black displayed slides warning that the burning of fossil fuels could eventually endanger humanity. "Present thinking holds that man has a time window of five to ten years before the need for hard decisions regarding changes in energy strategies might become critical," the scientist later summarized in a memo. Black had also identified the prime perpetrators, describing "general scientific agreement that the most likely manner in which mankind is influencing the global climate is through carbon dioxide release from the burning of fossil fuels."

That memo wouldn't surface publicly for almost 40 years, when the Pulitzer Prize-winning nonprofit newsletter, *Inside Climate News*, revealed the discovery of these early dire warnings in ExxonMobil's own archive. This was followed, in fall 2015, by an exposé published in the *Los Angeles Times*. The *Times'* reporters, plus a team of Columbia University post-graduate journalists, pieced together the damning saga primarily from hundreds of documents housed in the ExxonMobil Historical Collection at the University of Texas, Austin. (A book, *Exxon: The Road Not Taken*, has subsequently been published.)

Further evidence of the cover-up surfaced in April 2016, in a report by *DeSmog Blog* based on corporate documents found in the archive of Exxon's Canadian subsidiary, Imperial Oil. A "Review of Environmental Protection Activities for 1978–1979" prepared by Imperial Oil stated there was "no doubt that increases in fossil fuel usage" were "aggravating the potential problem of increased CO_2 in the atmosphere. Technology exists to remove CO_2 from stack gases but removal of only 50 percent of the CO_2 would double the cost of power generation." Managers all through Exxon's international offices were on the distribution list for this alarming report. A subsequent company report for 1980–81 noted as one of the "Key Environmental Issues and Concerns" that global warming is "receiving increased media attention."

Exxon wasn't alone in its malfeasance. Back in 1979, the American Petroleum Institute established an industry task force to monitor and share research on climate impacts. The members of the joint project included senior scientists and engineers from Exxon and nine other energy firms—Amoco, Phillips, Texaco, Shell, Mobil, Sunoco, Sohio, Gulf Oil and Standard Oil of California. A background paper produced for the group stated that atmospheric carbon dioxide was steadily accelerating. Initially calling itself the CO_2 and Climate Task Force, the group changed its name to the Climate and Energy Task Force in 1980. "It was a fact-finding task force," according to former director James J. Nelson. "We wanted to look at emerging science, the implications of it and where improvements could be made, if possible, to reduce emissions."

During a meeting of the task force in February 1980, a Texaco representative proposed that the group should "help develop ground rules for energy release of fuels and the cleanup of fuels as they relate to CO_2 creation." The following year, Exxon hired Harvard astrophysicist Brian Flannery as an in-house scientist to specifically conduct research into the impact of greenhouse gas emissions. That spring, Flannery sat on a Department of Energy (DOE) workshop panel alongside NASA scientist James Hansen, who would shortly publish a study in the journal *Science* warning about significant warming—even *if* emissions controls got put in place. "Scientists are agreed," the workshop concluded, that an ongoing atmospheric buildup would pose problems for the biosphere.

Exxon scientists had by now outfitted an oil tanker with CO_2 detectors and analyzers, while constructing models to project how global temperatures would be affected by a doubling of greenhouse gases in the atmosphere. A 1982 corporate primer produced by the company's environmental affairs office recognized that "major reductions in fossil fuel combustion" would be required; otherwise, "there are some potentially catastrophic events that must be considered. Once the effects are measurable, they might not be reversible." Exxon marked this disturbing document "not to be distributed externally"—even though it contained information that "has been given wide circulation to Exxon management."

Despite the growing awareness of the carbon dioxide crisis within Exxon's managerial circles, a June 1982 corporate memo noted that CO_2 program expenditures should be trimmed from $900,000 a year to no more than $150,000, starting immediately. (At the time, Exxon's annual research-and-development budget topped $600 million, while its exploration and capital budgets stood at $11 billion.) Around the same time, a corporate study acknowledged that "the increasing level of atmospheric CO_2 is causing considerable concern due to potential climate effects," but added that an expanded research effort "would require skills which are in limited supply." Two innovative experiments were soon terminated—one on the oceans' ability to absorb CO_2 and another to test vintage French wines for telltale traces of the greenhouse gas.

Martin Hoffert, then a physics professor at New York University, joined Exxon as a consultant in the early 1980s. A decade earlier, Hoffert had been a senior research fellow at NASA's Goddard Institute for Space Studies, working alongside climate science pioneers James Hansen and Stephen Schneider. Climate modeling was, Hoffert recalls, "a very esoteric field back then, because although scientists had hypothesized at the end of the 19th century that climate would change from burning fossil fuels, the temperature of the earth was cooling in the 1970s. We were basically geeks who started writing computer simulations and writing papers. I did a back-of-the-envelope calculation that, at the rate carbon dioxide was going up into the atmosphere, by the end of the 1980s it would become the dominant factor in climate change."

Hoffert would become the author or co-author of a majority of Exxon's nearly 50 peer-reviewed papers on the topic. In a highly technical 50-page chapter co-written by Hoffert and Brian Flannery for a 1985 Department of Energy report, the scientists predicted that, unless emissions were scaled back, the start of the 21st century could witness a staggering temperature rise of up to six degrees Celsius. Hoffert says he "very naively believed that the signal would break through and the scientists would tell politicians to do what was necessary. At that time, there were no divisions, no agendas. We were coming together as scientists to address issues of vital importance to the world."

However, "there was a fork in the road. They had the opportunity to make a decision to go one way or the other way. If Exxon had listened to its scientists and endorsed our research . . . it would have had, in my opinion, an enormous impact." Instead, "what happened was an incredible disconnect in people trained in physical science and engineering. It's an untold story of how we got to the point where climate change has become a threat to the world."

The summer of 1988 was a memorable one, a turning point in terms of increased awareness about global warming. A scorching heat wave was blamed for more than 5,000 deaths in the US and costs of close to $40 billion. Hoffert's former NASA colleague, James Hansen, issued a warning before Congress about the consequences of failing to act. But meanwhile, Exxon was already substituting spin for science, with its public affairs manager recommending, in an August 1988 internal memo, that the company "emphasize the uncertainty" in the scientific data concerning the role of fossil fuels.

Corporate paranoia was running high at the time. On March 24, 1989, the tanker *Exxon Valdez* had run aground on a reef in Alaska's Prince William Sound. Almost 11 million gallons of crude oil spewed from a ruptured hull into the remote, pristine waters—the largest oil spill in American coastal waters until BP's *Deepwater Horizon* disaster 21 years later.

Around the time of the *Valdez* spill, Duane LeVine, Exxon's manager of science and strategy development, gave a presentation to the company's board of directors. "Data confirm that greenhouse gases are increasing in the atmosphere. Fossil fuels contribute most of the CO_2," LeVine said unequivocally. By the middle of the 21st-century, global temperatures would likely rise between 2.7 and 8.1 degrees Fahrenheit, bringing melting glaciers, rising sea levels and "generally negative consequences." Already, pressure from environmentalists was mounting, LeVine said, citing the recent Montreal Protocol that banned ozone layer-depleting chlorofluorocarbons (CFCs) and adding that this "pales by comparison to the difficulties of applying similar approaches" to carbon dioxide. LeVine went on: "Arguments that we can't tolerate delay and must act now can lead to irreversible and costly draconian steps."

Scientist Flannery weighed in with a note to colleagues in an internal Exxon newsletter, warning that regulatory attempts to reduce risk would "alter profoundly the strategic direction of the energy industry." At the annual meeting of its shareholders in 1990, a proposal calling upon Exxon to reduce emissions was denounced by the company's board due to the "great scientific uncertainties."

What had happened to company scientists like Flannery? "Brian knows everything I know, he was a very smart guy," Hoffert says. "But I think he and others began to feel the weight of the front office. You're doing all this research, but you have a nice house in the suburbs. Do our brains work in such a way that we adopt the feelings and belief systems of people in our tribes? You don't want to lose the community that validates you. I did talk to them [the Exxon scientists] about it all the time when I was there, but they would just say I was a liberal university guy. And why didn't the higher-ups at Exxon feel ethically disturbed by what they were learning? Maybe some did; it's possible some execs quit or voted against their bosses at the board meetings, but we'll never know."

As Hoffert wrote from retirement in a private essay, "My Exxon History of Climate Research," in 2015: "Frankly I'm not sure if I quit or was fired as one of their major consultants on climate change science. . . . Why, I wondered forty years back, couldn't a giant multinational hydrocarbon company like Exxon redefine and reconfigure itself as an energy company for the twenty-first century, much like General Electric and even Silicon Valley based companies like Tesla Motors are actually doing?. . . . This path not taken here can cost our children and grandchildren dearly. The details of this story, still in progress, need to be told, analyzed, debated and eventually shouted from the rooftops."

Hoffert's own rooftop in central Florida, designed himself, is lined with $50,000 worth of photovoltaic solar cells, making his a nearly carbon-neutral residence. The retired scientist drives a plug-in hybrid and exercises on a solar-powered bicycle.

Lee Raymond, dubbed "Iron Ass" by company insiders, was manning the corporate ship at the time Hoffert got dismissed. He would turn Exxon into a leading player in the climate change denial movement. A native South Dakotan who went to work for Exxon in 1963, Raymond had worked his way onto the Exxon board within two decades, where his responsibilities included oversight of the Exxon Research and Engineering department that Hoffert served. In 1987, Raymond became Exxon's president and in 1993, he also became CEO —the same year the company left the New York headquarters where founder John D. Rockefeller had set up shop more than a hundred years before, and shifted its domain to Dallas. At the time of Raymond's ascension, Exxon was by far the largest oil company in America, twice as big as Mobil Oil and bigger still than Chevron—both progeny as well of Rockefeller's original Standard Oil monopoly. Raymond, who also sat on the board of the J.P. Morgan investment bank, thought the words "crude oil" ought to be carved in stone outside Exxon's Texas headquarters.

Not long after the first meeting of the United Nations' Intergovernmental Panel on Climate Change (IPCC) in 1989, a new Global Climate Coalition (GCC) formed in Washington. It operated out of the office of the National Association of Manufacturers, and described itself as "an organization of trade associations established . . . to coordinate business participation in the international policy debate on the issue of global climate change and global warming." Early members included Exxon, Chevron, Shell Oil, Amoco, Texaco, General Motors, Ford, Chrysler, the American Forest & Paper Association, AMAX Minerals, and the US Chamber of Commerce. Serving as chair was the American Petroleum Institute's then-executive vice president, William O'Keefe.

The GCC employed PR veterans of earlier battles against environmentalists—including the company that, on behalf of the pesticide industry, had spearheaded an attack in the 1960s against Rachel Carson's environmental classic, *Silent Spring*. In the months prior to the Earth Summit at Rio de Janeiro in 1992, the GCC led a successful lobbying effort to keep the US from endorsing mandatory emissions controls. It also provided a video to journalists at the summit, turning the whole argument upside-down: World hunger, claimed the GCC, could actually be alleviated by *higher* levels of CO_2 because these increased crop production.

The Global Climate Coalition was busy on many fronts, producing clouds of disinformation and confusion about the growing crisis. It coordinated with the National Coal Association to spend over $700,000 on the climate issue in 1992 and 1993, and with the American Petroleum Institute on a "grassroots" letter and telephone campaign to prevent a proposed tax on fossil fuels (the API paid $1.8 million to the Burson-Marsteller PR firm in 1993).

According to the Center for Media & Democracy's *SourceWatch*: "The GCC website was decorated with numerous photos of happy children playing in idyllic farm fields, but it did not provide any information about its budget or where its money comes from. GCC was not registered as a nonprofit organization and was not required to make public disclosures of its IRS tax filings, so it is difficult to obtain even basic information about its finances. However, the information that is publicly available shows that the GCC has spent tens of millions of dollars on the global warming issue."

The industry escalated its climate change propaganda battle throughout the years of Bill Clinton's presidency, with a think tank called the Competitive Enterprise Institute receiving hundreds of thousands of dollars from Exxon's public affairs shop, enabling CEI to file lawsuits challenging "unreliable" climate studies. At the same time, GCC pumped millions into a campaign specifically aimed at derailing the Kyoto Protocol, a global treaty designed to require emissions reductions by the wealthy industrialized nations. The organization held a "Costs of Kyoto" conference and spent over $3 million on newspaper and TV ads claiming that a "50-cent-per-gallon gasoline tax" would emerge if the treaty succeeded.

The Global Climate Coalition and similar industry-funded groups succeeded in raising doubt among Americans. While a 1992 poll found that 88 percent of Americans believed global warming a serious problem, by 1997, the number had fallen to 42 percent, with only 28 percent thinking any immediate action was needed.

Early in 1997, Exxon CEO Raymond flew to Beijing to address the yearly gathering of the World Petroleum Congress. The Clinton administration was then in the final round of negotiations on Kyoto, and Raymond had one purpose in mind: attack the agreement as unnecessary. "Natural fluctuations" in temperature had occurred all through history, Raymond said. Fully 96 percent of the CO_2 entering the atmosphere came from nature, Raymond claimed. Astoundingly, Raymond urged the Chinese government to defy the democracies of the US and Europe and block an agreement that would cause "slower economic growth, lost jobs and profound and unpleasant impact on the way we live."

In 1998, Exxon took the lead in creating a new Global Climate Science Team, including the company's chief lobbyist Randy Randol and a PR representative for the American Petroleum Institute. The team's initial memo, leaked to the *New York Times*, said: "Victory will be achieved when average citizens 'understand' (recognize) uncertainties in climate science" and when that "recognition of uncertainty becomes part of the 'conventional wisdom.'" Putting the plan in place was William O'Keefe of the GCC, by now president of the George C. Marshall Institute think tank but simultaneously doubling as a lobbyist for Exxon.

As his speech in Beijing indicated, Lee Raymond felt no need to concern himself with the public good or political will of the country where he was born. "I'm not a US company and I don't make decisions based on what's good for the US," he boldly stated. "Presidents come and go. Exxon doesn't come and go." At the end of 1998, as if to reinforce that Exxon was a colossus that superseded national governments, Raymond made a $75 billion deal to merge with Mobil, Exxon's second biggest rival.

Although the Kyoto Protocol marked the first time that nearly all Western nations agreed to cut emissions, the US never did ratify the treaty. Neither did China. In one of his early acts as president, George W. Bush pulled the US out of the protocol altogether on March 28, 2001, saying he wouldn't do anything to "harm our economy and hurt our workers." ExxonMobil would claim it had never "campaigned with the United States government or any other government to take any sort of position over Kyoto." But Under Secretary of State Paula Dobriansky would tell the Global Climate Coalition, of which the company was a leading member: "POTUS [the president of the United States] rejected Kyoto, in part, based on input from you." According to Neela Banerjee of *Inside Climate News*, "During the Bush administration, some of the key people involved in the Global Climate Coalition went on to . . . top administrative posts, and some of them worked to censor science on climate change."

Lee Raymond clearly played a key role in moving Bush away from his earlier campaign position calling for new limits on how much carbon dioxide could be emitted from power plants. The ExxonMobil chief and Vice President Dick Cheney—former CEO of Halliburton, the oil-services and infrastructure giant—had known each other "very well" (in Raymond's words) for over two decades. They'd gone hunting together when Cheney was a US Congressman from Wyoming. They lived close to one another in Dallas after Exxon moved its headquarters there. Exxon was a major Halliburton client throughout Cheney's tenure at the company. And Raymond sat on the board of the American Enterpirse Institute, where Cheney's wife Lynne was a senior fellow. Unsurprisingly, the conservative think tank was a bastion of anti-regulatory opinion when it came to climate change.

On January 29, 2001, nine days after Bush's inauguration, Raymond came to the West Wing to meet with his pal Cheney. The new vice president had already established an energy task force, whose private policy meetings with industry executives and lobbyists the administration would keep secret. Bush had simultaneously formed a cabinet-level group to review climate science and policy. Some of its members later opined that the president sincerely believed that we shouldn't keep postponing regulation of greenhouse gases. That was before Cheney reportedly applied pressure on the president, soon pushing Bush to sign a letter to Congress repudiating his campaign rhetoric. According to author Steve Coll in his book *Private Empire: ExxonMobil and American Power*, Bush never bothered to inform his EPA chief Christine Todd Whitman about his backpedaling on the climate issue. She reportedly called Treasury Secretary Paul O'Neill and said: "Energy production is all that matters. [Cheney] couldn't have been clearer." O'Neill is said to have replied: "We just gave away the environment."

With Bush and Cheney now "the deciders," ExxonMobil had political friends in the highest places. But the oil giant did not relent on the propaganda front.While openly supporting the larger climate-change-denier front groups like the American Enterprise Institute, Heritage Foundation and Cato Institute, ExxonMobil also secretly funneled contributions to "small, havoc-making groups" that published books like *The Global-Warming Deception: How a Secret Elite Plans to Bankrupt America and Steal Your Freedom*. Raymond also developed a slide show depicting thousands of scientists as doubters.

But all was not serene within the energy giant's corporate suites. Frank Sprow, the ExxonMobil vice-president in charge of safety and environment, reportedly confronted Raymond, telling the CEO, "Is it not the case that the risk of climate change is high enough that responsible efforts . . . to mitigate risk would be worthwhile?" In 2002, the Global Climate Coalition disbanded and ExxonMobil instead poured money into financing a Global Climate & Energy Project at Stanford, giving $100 million of a $225 million total also contributed by General Electric, Toyota and Schlumberger (the world's biggest oil services company). The research was allegedly designed to seek breakthroughs in clean energy technologies.

The year 2002 found Martin Hoffert, no longer a consultant, as lead author on a paper published in the prestigious *Science* journal titled

"Advanced Technology Paths to Global Climate Stability." Hoffert wanted to get ExxonMobil officials to support his research paper, which addressed alternatives to fossil fuels. "My motivation was [in part] to get ExxonMobil to sign on as one of the authors. I thought this could be important in combating the partisanship crippling action on the climate consensus.... Down to the final galley proof submission deadline, I was negotiating the wording of the paper to keep an ExxonMobil author onboard."

Exxon researcher Haroon Kheshgi's name does appear on the study, but Hoffert says Kheshgi had to get front-office clearance and "they kept suggesting editorial changes, everything that made it look like there was no controversy about climate change. At the last minute, I changed only one word. But I know that Exxon's managers were reading it very carefully for implications, because I guess they knew somebody might look in the future."

State Department papers from June 2005 show the Bush administration thanking ExxonMobil executives for the corporation's "active involvement" in shaping climate change policy, which remained do-nothing. Lee Raymond told a reporter: "We think we have a responsibility. If we think people are about to make some bad policy decisions that are going to have a big impact for a long period of time, somebody's got to say something."

In 2006, the prestigious Royal Society of the United Kingdom issued a powerful criticism of the oil company for issuing "very misleading" statements about the IPCC's Third Assessment Report, to which one of Exxon's own scientists had contributed. The society's communications manager declared that Exxon funded at least 39 organizations "featuring information on their websites that misrepresent the science on climate change."

Lee Raymond was about to turn 65. In his final year as CEO, ExxonMobil earned the largest net profit of any corporation in history, $36.1 billion. Raymond took a retirement package valued at close to $400 million and moved to the Dallas suburb of Westlake, where some 700 residents occupy America's most affluent neighborhood.

In January 2006, Raymond turned over ExxonMobil's corporate reins to 53-year-old Rex Tillerson, who'd cut his teeth in the exploration division. At the same time, ExxonMobil scientists involved in locating new oil and gas deposits began assessing how the work of acclaimed earth scientist Peter Vail might lead to new discoveries, if climate change did in fact alter sea levels and land surfaces. "So don't believe for a minute that ExxonMobil doesn't think climate change is real," said an anonymous former company manager interviewed by Steve Coll. "They were using climate change as a source of insight into exploration."

Tillerson, who's called Ayn Rand's *Atlas Shrugged* his favorite book, served as Raymond's executive assistant early in his career, but did not have his former's boss's combative style. Raymond believed that, since its inception under the Rockefellers, the company had a history of standing tough when it had to. He reportedly screened Al Gore's Oscar-winning documentary, *An Inconvenient Truth*, numerous times, taking notes and assembling talking points against Gore's argument that climate change posed a threat. "The scientists on the other side are wrong," Raymond asserted flatly.

Tillerson, too, refused to accept the scientific consensus that there was indeed a cause-and-effect link between carbon emissions and rising temperatures. But, unlike Raymond, he took a more subtle approach to countering the company's environmental critics, saying, "It's more complicated than most people understand." Yes, it was prudent to put in place strategies that addressed the risk, he acknowledged, while always "keeping in mind the importance of energy to the economies of the world." Public health, economic development, and the eradication of poverty should be paramount, Tillerson emphasized.

Raymond's successor seemed to think the climate change battle was a game whose clock had to be managed, giving the company enough time to exploit its enormous investment in fossil fuels until it was finally forced to make a full transition to renewable energy. Tillerson did concede that eventual battery breakthroughs and biofuels, perhaps solar and wind energy, would change the energy economy. But, as ExxonMobil's in-house climate management committee attested in 2008, the company could feel secure about its oil and gas investments for a good two decades and likely much longer.

Shortly after Barack Obama won the presidency, Tillerson traveled to Washington to reveal ExxonMobil's new, more sophisticated lobbying strategy in a speech at the Woodrow Wilson International Center for Scholars. While climate change was an "important global issue," Tillerson conceded, he emphasized that "there is enough oil and natural gas offshore and in non-wilderness and non-park lands to fuel fifty million cars and heat nearly one hundred million homes for the next twenty-five years." Though he didn't entirely reject the Obama adminstration's call for a cap-and-trade system to limit carbon emissions, Tillerson offered a tax alternative that he knew was a political non-starter. After he discussed his idea with Obama's Environmental Protection Agency chief Carol Browner, she said privately that Tillerson "was happy to have a position that nobody was going to embrace."

Here's the position that Tillerson *did* embrace: Exxon's 2008 Action Plan stated the need to "identify, recruit and train a team of five independent scientists to participate in media outreach. These will be individuals who do not have . . . visibility . . . in the climate change debate. Rather, this team will consist of new faces." The shiniest of those faces belonged to Willie Soon and Sallie Baliunas, both connected to the Harvard-Smithsonian Center for Astrophysics. "I would never be motivated by money for anything," Soon assured the *New York Times*, after proclaiming that variations in the sun's energy are behind global warming. In fact, Soon had received more than $1.2 million from ExxonMobil, the American Petroleum Institute, Southern Company (a utility holding company vested in coal-burning power plants) and the Charles G. Koch Charitable Foundation. Soon never disclosed these ties in the 11 climate change papers he published after 2008, at least eight of which appear "to have violated ethical guidelines of the journals." Documents obtained by Greenpeace under the Freedom of Information Act included correspondence that Soon sent to corporate funders calling many of his papers "deliverables" in exchange for cash. Soon, actually a part-time employee of the Smithsonian with a Ph.D. in aerospace engineering, has almost no formal training in the subject of climatology.

ExxonMobil's profits continued to soar under Tillerson. In 2010, while receiving an award in Houston for contributing to the city's international profile, the CEO was asked a question about solar and wind power. Tillerson replied, "ExxonMobil is not really against renewables. We sell a lot of lubricant to the windmill operators . . . the more windmills are built, the more oil we sell." The very next day, BP's oil spill disaster occurred in the Gulf of Mexico. The accident, Tillerson claimed, was "a dramatic departure from the industry norm in deep-water drilling." ExxonMobil, he said, would never have made such a mistake.

Despite Tillerson's softer approach, his company continued to throw its weight behind the anti-science disinformation campaign. Since 2007, Exxon has contributed $1.8 million to the campaign coffers of over 100 members of Congress who deny climate change is happening. Those friends on Capitol Hill have successfully ensured that no carbon tax bill has been passed. Between 1998 and 2014, the ExxonMobil Foundation paid out at least $22 million dollars to as many as 100 think tanks and other research-for-hire organizations that ran climate denial campaigns. Recipients included the Advancement of Sound Science Coalition, which first made its mark challenging the accepted medical evidence on the hazards of second-hand smoke. Others funded were the Heartland Institute, the George Marshall Institute, the Committee for a Constructive Tomorrow, and the American Legislative Exchange Council, which obstructs renewable energy initiatives in numerous state legislatures. The Competitive Enterprise Institute, after receiving $90,000 in "General Operating Support" in 2005, produced a minute-long TV commercial headlined "Carbon Dioxide: They Call it Pollution, We Call it Life." Its release brought such a strong public outcry that afterwards ExxonMobil ceased funding the organization.

Tillerson, in June 2012, delivered an address to the Council on Foreign Relations in New York, announcing that the US was about to achieve "energy security"—thanks, in large part, to the growth of fracking. Tillerson, whose company had become the largest domestic producer of natural gas after acquiring XTO Energy, also had strong words for his critics: "Ours is an industry that is built on technology, it's built on science, it's built on engineering, and because we have a society that by and large is illiterate in these areas—science, math and engineering—what we do is a mystery to them, and they find it scary. And because of that, it creates easy opportunities for opponents of development—activist organizations—to manufacture fear . . . that's how you slow this down. And nowhere is it more effective than in the United States."

It was a remarkably brazen line of attack for a multinational corporation that had taken the lead in financing anti-science propaganda for four decades.

Tillerson went on, in response to a question about the environmental impacts of burning fossil fuels: "I'm not disputing that increasing CO_2 emissions in the atmosphere is going to have an impact. It'll have a warming impact. . . . [But] as human beings . . . we have spent our entire existence adapting, okay?. . . . Changes to weather patterns that move crop production areas around—we'll adapt to that. It's an engineering problem, and it has engineering solutions."

This was shortly before Hurricane Sandy devastated the Eastern seaboard of the US, one more climate calamity tied to the earth's unsustainable levels of carbon buildup.

At a May 2013 meeting of the company's shareholders in Dallas, Tillerson sounded less sanguine about the ability of good old American know-how's ability to engineer the world out of the climate-change crisis: "We do not see a viable pathway with any known technology to achieve [CO_2 reduction] that is not devastating to economies, societies and people's health and well-being around the world. What good is it to save the planet if humanity suffers?"

The day that the IPCC's Fifth Assessment Report in 2014 confirmed that climate change is impacting every part of the globe, ExxonMobil decided to release a report of its own. This was the first Carbon Asset Risk study by a major oil and gas producer, requested by certain shareholders. Yes, said Exxon spokesman Kenneth Cohen, the company now knew enough "based on the research and science that the risk [of climate change] is real and appropriate steps should be taken to address that risk. But given the essential role that energy plays in everyone's lives, those steps need to be taken in context with other realities we face, including lifting much of the world's population out of poverty." After all, our world would require 35 percent more energy in 2040 than in 2010. New forms of energy couldn't possibly supplant the status quo, and it was "highly unlikely" that ExxonMobil would cease selling fossil fuels for many decades to come.

The Exxon strategy had clearly pivoted to embracing the science, but dragging its feet on the solutions, as long as there were billions more in profit to be made from extracting fossil fuels from the land, sea or melting glaciers.

Climate change wasn't even mentioned in an early 2015 report by the National Petroleum Council's Arctic Committee, which Tillerson chaired. No mention that 2014 (which saw ExxonMobil earn $32 billion) had been the hottest year ever recorded, or that the Arctic region was melting faster than anyone could have anticipated. Rather, the Petroleum Council concluded, Arctic drilling for oil and gas needed to move forward. A study published in the journal *Nature* concluded, in dissent: "ExxonMobil may spend tens of billions to develop Arctic reserves that can't safely be used."

But Exxon saw the environmental destruction of the Arctic, for which it shared major responsibility, as a new business opportunity. Years ago, Exxon began leasing vast tracts of the region for future oil exploration, because "potential global warming can only help lower exploration and development costs," as a company scientist noted in a

1990 memo. Indeed, the energy giant began "climate-proofing" on other fronts: raising the decks of offshore platforms to combat sea level rise and protecting pipelines from increased erosion, from the Canadian Arctic to the North Sea.

In 2011 Exxon had signed a lucrative deal to explore Russia's Arctic Kara Sea with the country's leading government-owned petroleum company, Rosneft. Its boss, Igor Sechin, is probably the Kremlin's second most powerful man after Vladimir Putin, who two years later awarded Tillerson the Russian Order of Friendship. After President Obama accused Russia of invading Ukraine and annexing Crimea in 2014, he imposed sanctions that forced Exxon to pull back from the prospective drilling project—resulting in a loss of as much as a billion dollars, according to a company regulatory filing.

In May 2015, Tillerson reverted to an old tactic in his speech to the annual meeting of ExxonMobil shareholders: he didn't even bring up climate change, and during the question and answer period, he dredged up the old line that more solid science was needed. "What if . . . it turns out our models are lousy, and we don't get the effects we predict?" he asked rhetorically. In the latest company report on the implications of its carbon footprint, ExxonMobil asserted that its energy outlook process doesn't project overall greenhouse gas concentrations nor does it model global average temperature impacts. A footnote added: "These would require data inputs that are well beyond our company's ability to reasonably measure or verify."

That hadn't been the case when its own scientists projected the terrifying future ahead, more than a generation earlier. It was as if Tillerson had come full circle to where Exxon had begun: the science was uncertain, more research needed to be done, and in the meantime, drill, baby, drill. In the course of the Paris climate negotiations, Exxon refused to add its support to a letter that other oil giants, including BP and Royal Dutch Shell, had signed urging a global price on carbon. Instead, its corporate statement on the Paris accord urged reduction of emissions "at the lowest cost to society, keeping in mind that access to affordable and reliable energy is critical to economic growth and improved standards of living worldwide."

When the story broke in autumn 2015 about ExxonMobil's research almost four decades earlier on the looming climate crisis, Tillerson tried to put the revelations in the best possible light, as if the company was ahead of its time in sounding the alarm, instead of guilty of a massive coverup. "We were spending a lot of time trying to understand this issue in the early days," he told Fox Business News. "We were very open with the work we were doing. Most of it was done in collaboration with academic institutions and many government agencies for us to understand this better. And I think as we began to understand that, and people began to think about policy choices, we had a view on policy choices which has not changed very much over the years, and we've been very open about that."

New York Attorney General Eric Schneiderman has grave questions about that professed openness. In fact, the company's failure to act on its early knowledge of the climate crisis might be grounds for criminal prosecution. Early in November 2015, ExxonMobil received a subpoena demanding financial records and emails dating back at least a decade. The New York investigation actually began a year earlier, after Harvard scientist Willie Soon was discovered to have received more than a million dollars from ExxonMobil, among others, to downplay climate problems. The New York state attorney general's investigation is seeking to determine whether ExxonMobil and other energy corporations committed financial fraud in deceiving the public and their shareholders about the risks of climate change.

The reasoning is similar to when tobacco companies in the 1950s and 1960s financed internal research that privately showed their product to be harmful and addictive, then initiated a public campaign including falsified scientific research to indicate precisely the opposite. As one tobacco executive put it in a memo in 1969, mirroring the climate skeptics of the future: "Doubt is our product, since it is the best means of competing with the 'body of fact' that exists in the minds of the general public. It is also the means of establishing a controversy."

The US Justice Department's suit against Big Tobacco led to the largest civil litigation settlement ever in 1998, with the companies agreeing to pay billions in fines to state governments for "a massive 50-year scheme to defraud the public." Sharon Eubanks, who prosecuted the racketeering case against the industry, now believes that ExxonMobil could be liable under the same statute for suppressing its knowledge about climate change. "Exxon-Mobil is not alone," emphasized professor Stephen Zamora of the University of Houston Law Center. "This is not likely to be an isolated matter."

The corporation quickly fought back. In 2014, the ExxonMobil Foundation had provided almost $200,000 to Columbia University as part of a matching gift program for educational institutions. Now ExxonMobil addressed a letter to Columbia's president and trustees. The company accused Steve Coll, dean of the vaunted Columbia School of Journalism and author of *Private Empire,* as well as a team of recent graduate students who examined the ExxonMobil archives on climate research, of violating the school's policy on research misconduct by downplaying or ignoring information provided by ExxonMobil itself. "The interactions . . . are not typical of the high standards and ethical behavior we have come to expect from your institution," corporate public affairs chief Kenneth Cohen wrote the university.

Ironically, the Columbia University investigation into Exxon's climate archives had been partly funded by a grant from the Rockefeller Brothers Fund. Cohen lambasted the fund as well, which had not only supported activism against the Keystone pipeline but vowed to divest from fossil fuels, for holding "a bias against the oil and gas industry."

The public affairs chief then appeared on a PBS News Hour interview, during which host Judy Woodruff asked whether ExxonMobil funded groups denying climate science. "Well, the answer is yes. And I will let those organizations respond for themselves," Cohen replied.

Eight years earlier, Cohen claimed the company had ceased all such funding. But perhaps it had been merely better disguised. According to the Climate Investigations Center, a Washington-based nonprofit, as of 2005, "things got strange. The public version of Cohen's ExxonMobil Foundation's grants contained no descriptions—instead vague, anodyne explanations (e.g. 'General Support'), whereas the forms the Foundation submitted to the IRS contained more detail about the grants. The public version is published in ExxonMobil's Worldwide Giving Report, released each spring around the annual shareholders meeting, and officially filed with the IRS as a '990' form. The 2005 ExxonMobil Foundation 990 lists a total of $996,500 in grants described as specifically for climate change-related work. The 2005 Worldwide Giving Report lists none." Among the foundation's recipients were the denier specialists Competitive Enterprise Institute and ALEC.

So it goes. Meanwhile ExxonMobil spends about $4 million a day looking for new oil and gas sources—new vast repositories of fossil fuels to be burned and turned into carbon waste—and Rex Tillerson continued receiving a salary of almost $100,000 a day for his single-minded focus on drilling and fracking.

But all is not well in the ExxonMobil empire. In 2015, Exxon's profits were the lowest in 13 years, and because of a debt that more than tripled, the corporation's top-ranked credit score disappeared. Then, early in 2016, the US Justice Department asked the FBI's Criminal Investigation Division to investigate whether ExxonMobil's years of funding climate deniers despite its knowledge of the truth constituted a potential violation of federal laws. Late in March, the state attorney generals from Massachusetts and the Virgin Islands joined New York Attorney General Eric Schneidermann and former Vice President Al Gore at a news conference, saying they too would investigate whether ExxonMobil had deceived investors and the general public about the risks posed by climate change. Gore, making a comparison with the Clinton administration's action against Big Tobacco, said: "I do think the analogy may hold up rather precisely."

Also early in 2016, the New York State Comptroller and four other ExxonMobil shareholders issued another challenge—asking the federal

Securities and Exchange Commission (SEC) to force the corporation to address how its bottom line "will be impacted by the global effort to reduce greenhouse gas emissions and what the company plans to do about it." ExxonMobil's response? Because the recent international agreement in Paris was a "practical improbability" in terms of achieving its goals, it need not bother to examine the impact the proposed regulations would have on its business. The company, as one reporter put it, "seems to have flipped from willful ignorance to purposeful complacency."

But it was getting harder to maintain its posture of nonchalance. In March, the SEC ruled that ExxonMobil had to include a climate change resolution when the annual meeting of shareholders came around again in May. The SEC rejected the company's claim that it already offers adequate disclosures on how its carbon output affects the future bottom line. "Investors need to know if ExxonMobil is taking necessary steps to prepare for a lower carbon future, particularly now in the wake of the Paris agreement," said New York State Comptroller Thomas DiNapoli.

When over 38 percent of the company's investors then voted at the May meeting to compel ExxonMobil to publish a yearly study concerning how climate change policies could affect its profits, Tillerson took the stage with his rebuttal. Be assured, he told shareholders, that Exxon had put more than $7 billion into green technology, but the breakthroughs still hadn't happened. "Until we have those," Tillerson went on, "just saying 'turn the taps off' is not acceptable to humanity. The world is going to have to continue using fossil fuels, whether they like it or not."

Tillerson received powerful backup from US Representative Lamar Smith, the Texas Republican chairman of the climate change-denying House Committee on Science, Space and Technology, who's raked in over $687,000 in donations from fossil fuel companies. Accusing the attorneys general of New York and Massachusetts of a "form of extortion" against ExxonMobil, Smith issued subpoenas against them and eight advocacy groups including Greenpeace, Bill McKibben's 350.org, and the Rockefeller Family Fund. The attorneys general refused to open their files, calling this an "unprecedented" effort to interfere with investigations by state regulators.

Meantime, the science-rejection crowd took control of the Republican Party with the nomination of presidential candidate Donald Trump. The 2016 GOP nominee hired Jim Murphy as his national political director for the presidential campaign. Murphy had been managing partner and finally president of the DCI Group for ten years, a Washington lobbying firm that represented Exxon in its ongoing efforts to stymie regulations on greenhouse gases.

Then, after Exxon filed a countersuit seeking to stop the investigations by the New York and Massachusetts attorneys general, in October a Texas federal judge issued a discovery order for the prosecutors to submit to questioning by ExxonMobil lawyers concerning why they are probing the corporation. This, as defenders of the legal system pointed out, was without precedent—"allowing wrongdoers possessing sufficient resources to file lawsuits that paralyze law enforcement efforts aimed at protecting the public." ExxonMobil has also issued subpoenas demanding to see the records of the nonprofit groups opposing them, a clear violation of the First Amendment.

Back in 2010, Rex Tillerson had recently stepped down as president of the Boy Scouts of America when he gave a keynote speech to the organization. He spoke of the Boy Scouts' "clear mission statement: serve more youth . . . and make the leaders of tomorrow, millions of 'em."

Whether those "leaders of tomorrow" will have a habitable planet to live on remains an open question that Tillerson and ExxonMobil refuse to address.

In the summer of 2015, at the National Order of the Arrow Conference's Gathering of Eagles—which drew nearly 5,000 scouts to a Michigan State University arena—Tillerson spoke about the importance of inculcating young men with a sense of personal integrity. "People trust you. They count on you," he said. "Your personal integrity, once established and earned, people don't have to think about it. They know. They know you. They know you'll do the right thing every time."

Tillerson is still on the Boy Scouts of America's board, but his professed interest in the future of youth doesn't stop there. He's the education chair for the Business Roundtable, a powerful trade association of more than 200 CEOs that's lobbied against climate action but has been pushing for rigorous nationwide school standards under a program called Common Core. Bill Gates, through his charitable foundation, has pumped over $220 million into Common Core, whose stated goal is to ramp up the dismal performance of American school children. (As of the latest international test results, American 15-year-olds ranked well below most other industrialized nations in both math and reading).

Tillerson, at a panel discussion in Washington in 2014, said he views setting high education standards as a "business imperative." American schools, he went on, "have got to step up the performance level—or they're basically turning out defective products that have no future. Unfortunately, the defective products are human beings. So it's really serious. It's tragic. But that's where we find ourselves today."

After a story quoting Tillerson appeared in *Fortune Magazine*, veteran educator Carol Burris offered this response in an open letter: "Dear Mr. Tillerson, Please leave our children alone. We do not need you to develop them as products. . . . In a world in which your corporation has been declared a person, one might mistake human children for products to consume. . . . The curtain has once again been pulled back to expose Oz, and the resistance to the corporate reform agenda will grow. . . .

"Mommies and daddies don't forget."

A public school teacher and activist named Steven Singer added in a press release: "Education is not about turning children into widgets for big business. It is about readying children for life, and that includes so much more than the tiny and inhuman vision of people like Rex Tillerson. Hands off my students, my daughter, and my country!"

ExxonMobil's corporate media relations senior adviser, William F. Holbrook, weighed in with an email response, saying: "To underscore what we've been saying publicly for some time, all children, regardless of where they live, must receive the best education possible for the United States to remain competitive globally."

Education about the drastically changing climate, perhaps? The hypocrisy of the planet's largest publicly traded oil and gas company seems to know no bounds. And there is no end in sight.

When nominated by President-elect Trump to be his Secretary of State, the 65-year-old Tillerson had already scheduled his retirement from ExxonMobil for 2017. He took a $180 million retirement package in early January, and was succeeded by Darren Woods, the man in charge of the company's worldwide array of oil refineries and fuel terminals. Like Tillerson, he will be both CEO and board chairman. A native of Wichita, Kansas, home of the Koch brothers, Woods will undoubtedly keep ExxonMobil focused on the extraction of fossil fuels.

According to *Bloomberg News*, "one of Woods's most-pressing tasks will be figuring out how to rescue a stillborn Russian joint venture that locked up $1 billion in investments and a billion-barrel Arctic oil discovery behind a wall of international sanctions." Before those sanctions were put in place by the Obama administration after Russia's invasion of the Ukraine, Tillerson had considered this deal his crowning achievement. Now he's been anointed America's top diplomat—the fox in charge of the global oil hen-house.

Tillerson's other big plan for Exxon, to develop the climate-polluting tar sands of Alberta, Canada, through the Keystone XL pipeline, is also likely to remain high on its agenda. Already, the company holds contracts for more than five billion barrels in Alberta, and ExxonMobil's website predicts that tar sands will be providing one-quarter of our oil by 2040. The website boasts: "The strength of our global organization allows us to explore across all geological and geographical environments, using industry-leading technology and capabilities." After all, at least 5,000 gigatons of carbon yet remain to be extracted.

In the mid-2000s, a group of scientists put their heads together on a "carbon budget" for earth. How much more fossil fuel could we burn before our planet headed on a trajectory of uninhabitable overheating? Maybe another 900 gigatons, they estimated.

It's ecocide. Simple as that. Burning down our own house. And despite the sci-fi fantasies of entrepreneurs like Elon Musk, there is no other planetary home for even the rich to fly to.

Do men like Tillerson and Raymond care that Earth might not be livable for future generations of their own families?

Like their father, Rex Tillerson's four children seem unperturbed by the fate of the planet. Two have followed in his footsteps to receive degrees at the University of Texas/Austin's Cockrell School of Engineering. The youngest boy is currently pursuing his mechanical engineering degree at the same institution. Meanwhile, Lee Raymond's only child, John T. Raymond, has held executive positions with numerous energy companies. Today, he is founder and majority owner of the Energy & Minerals Group (EMG), a private Houston-based equity firm, a company with an unfortunate track record in the fracking business.

But if these ExxonMobil heirs are oblivious to the grim environmental heritage their fathers have left behind, there are other descendants of fossil fuel wealth who have a deeper understanding of the grave prospects that now face humanity and what must be done to save life on the planet.

Chapter 3

The Rockefeller Family's Revolt Against ExxonMobil

It began in 2003. Neva Rockefeller Goodwin, a fourth-generation descendant of the Standard Oil founder and third child of banking mogul David Rockefeller, had recently become acquainted with Sister Pat Daly, a Boston activist leading an effort to influence corporate shareholders about the need to combat climate change. Goodwin and Daly agreed to co-sponsor a resolution at ExxonMobil's annual meeting to study the potential environmental impacts of the increasing carbon buildup (This was a dozen years before the corporation's early scientific research and then-existing knowledge of the onrushing dangers became public knowledge in 2015.)

Goodwin turned for support to two dozen Rockefeller first cousins. "Most members of our family will own shares of Exxon for far longer than the present management will be in place, and therefore we have an important interest in and responsibility toward the long-term viability of the company," five of these family members wrote in an email.

Although the shareholder resolution didn't pass, the Rockefeller family's concern about the climate issue was piqued. On two occasions in

2004, members met with investor-relations executives for luncheons at ExxonMobil's then-offices in Manhattan's Rockefeller Center. "We wanted to start talking with the company about their view of the future and how they could be a constructive player as well as part of the problem," recalled Goodwin, an economist and co-director of Tufts University's Global Development and Environment Institute. "The head of investor relations was really surprised we didn't love Exxon as it was but thought changes might be a good idea.".

The following year, numerous Rockefeller descendants came together for a private educational conference to educate themselves, inviting climate scientists, investors and industry executives. By the time Rex Tillerson took over ExxonMobil's helm from Lee Raymond early in 2006, an activist group of 43 family members had asked to meet with the CEO and the corporate board. Instead, Raymond and Tillerson chose to invite only David Rockefeller to a Manhattan lunch. The family patriarch—former chairman of Chase Manhattan Bank and over 90 years old at the time—had preached to his heirs that investor activism was "mostly carried out by nuts," according to Goodwin. Eventually, she received an invitation to the lunch as well, but later recalled not hearing "anything they haven't said other times to many people."

In 2016, she recalled of the meetings: "They were civil, but barely. . . . I believed, based on what I understood of the science, that the company's stance on climate change was simply incorrect; I had not assumed that they were lying."

Ongoing efforts to introduce shareholder resolutions on climate change continued to go nowhere. Then, in October 2006, West Virginia Democratic Senator Jay Rockefeller (a great-grandson of the Standard Oil founder), along with Republican Olympia Snowe of Maine, came forward with a strongly worded letter to Tillerson, criticizing the company's extensive funding of non-peer-reviewed pseudo-science. "Climate change denial has been so effective because the 'denial community' has mischaracterized the necessarily guarded language of serious scientific dialogue as vagueness and uncertainty," the senators wrote. "Exxon-

Mobil is responsible for much of this scientific data debate and support of global warming deniers." The corporation, Rockefeller and Snowe insisted, should make public such efforts and immediately cease its funding of the Competitive Enterprise Institute and organizations with similar purposes.

Tillerson didn't respond. A multinational corporation with the Olympian power of ExxonMobil doesn't feel compelled to acknowledge the complaints of mere US senators, even one with the lineage of Senator Rockefeller. The corporate board continued to refuse to meet with Rockefeller family members, who currently hold a very small percentage of the stock. The largest ExxonMobil shareholders are institutions, primarily those managing pension funds, and these investors customarily abstain on proposed resolutions

When the family gathered in 2007 for its yearly retreat at the original Rockefeller estate in the Pocantico Hills overlooking the Hudson, it zeroed in on separating Tillerson from being both CEO and board chairman of the company. More than 93 percent of the 78 direct adult descendants of John D. Rockefeller, the family later asserted, supported a proposal for an independent chairman. Although such so-called proxy resolutions are nonbinding, it was hoped that substantial shareholder support would be hard for ExxonMobil to ignore.

Under growing family pressure, the company tossed back a bone to the Rockefellers. Its 2007 Corporate Citizenship report promised that, the following year, "We will discontinue contributions to several public policy groups whose position on climate change could divert attention from the important discussion on how the world will secure energy required for economic growth in an environmentally responsible manner."

What did that carefully constructed sentence really mean? As an investigation by *The Guardian* revealed in 2015, rather than those "several public policy groups," ExxonMobil went on to contribute $1.7 million to Republicans in Congress who deny climate change (including Oklahoma Senator James Inhofe, who has called global warming a hoax) and another \$454,000 to the American Legislative Exchange Council (ALEC), the far-right lobbying group that has worked with state legislatures in suppressing alternative energy while hosting seminars extolling rising CO_2 emissions as an "elixir of life."

The company's ruse worked. Management's professed change of heart was enough to head off passage of the Rockefeller-supported resolutions at the 2008 shareholders meeting. Neva Goodwin did take the floor of the Dallas auditorium to explain that members of the extended Rockefeller family "have become seriously concerned about the future of Exxon. . . . Increased CO_2 in the atmosphere will cause weather disasters that . . . will certainly have a huge and harmful impact on the global economy itself, on which the success of a giant global company such as ExxonMobil so largely depends." But her resolution proposing ExxonMobil set up a task force to study climate change and develop sustainable energy technologies garnered only about 10 percent of the vote. The proposal to strip Tillerson of his board chairman title did better, receiving nearly 40 percent approval.

"I was pretty discouraged," Goodwin recalled. "Exxon has an extremely strong culture of believing that they are right and know what they are doing and really don't need to listen to anybody else. It was clear that we didn't have an ability to make more of a dent in that."

But the family didn't give up the fight. In 2011, Jay Rockefeller confronted Tillerson at a hearing of the Senate Finance Committee on tax breaks utilized by the big oil companies. When he accused the CEO of being "out of touch," Tillerson responded coolly that he wasn't "at all." Rockefeller fumed. "I think the main reason that you're out of touch, particularly with respect to Americans and the sacrifices that we're having to look at here in trying to balance, or come even close to balancing a budget, is that you never lose. . . . You always prevail. You always prevail in the halls of Congress and you do that for a whole variety of reasons, because of your lobbyists . . . because of all the places where you do business. And I don't really know any other business that never loses."

But in June 2014, ExxonMobil finally did suffer a loss—a highly symbolic setback that occurred, appropriately enough, on the grounds of the Rockefeller family estate. The Rockefeller Brothers Fund (RBF), a mega-philanthropic foundation established by heirs in the 1940s, had set out in the 1980s to support alternative energy, but without much success. That June, RBF board members escalated their efforts by voting to join a burgeoning movement around the country supported by local governments, religious organizations, pension funds and foundations. Modeled on the 1980s global campaign to combat apartheid by pushing multinational corporations to divest their holdings in South Africa, this new clean energy movement was aimed originally at universities with investments in the oil and coal industries. It was not surprising when the student-driven movement, sparked by author/activist Bill McKibben of the organization 350.org, began having success on college campuses, but when the Rockefeller Brothers Fund jumped on board the Big Oil divestment campaign, it marked a significant turning point in the battle.

The RBF board's vote took place on September 22, 2014, one day after 400,000 gathered for the People's Climate March down the streets of New York in the largest mass demonstration yet against the carbon polluters. By then, over 800 institutions and individuals, who together controlled over $50 billion in assets, had committed themselves to divest from the oil and coal industries. The RBF, half of whose grant money was already committed to fighting climate change, now became one of the biggest names in the growing climate change coalition.

"We all have a moral obligation," said Valerie Rockefeller Wayne, a great-great-granddaughter of the Standard Oil founder and chair of the RBF board. "Our family in particular—the money that is for our grant-making, and what we are doing now, and that helps fund our lifestyles came from dirty fuel sources."

RBF President Stephen Heintz said: "For a fund that is so devoted to fighting climate change and helping to prevent climate catastrophe, to continue to be invested in fossil fuels that are actually causing climate change just was morally hypocritical and unacceptable. . . . It felt like we were compromising ourselves."

With an endowment of more than $4 billion, the RBF would start by dumping its holdings in coal and tar sands, pulling out more slowly from oil and gas, and investing ten percent of its endowment in renewable sources. It was a start. As Wayne put it, "There is a moral imperative to preserve a healthy planet."

In January 2016, a conference took place around climate education efforts and activism at the New York offices that two Rockefeller family funds share. One topic they discussed, said a participant later, was "to establish in public's mind that Exxon is a corrupt institution that has pushed humanity (and all creation) toward climate chaos and grave harm."

Neva Rockefeller Goodwin wasn't done either. Early in 2016, she publicly announced in an op-ed essay published in the *Los Angeles Times* that she had donated all her inherited shares of ExxonMobil stock to the nonprofit Rockefeller Family Fund's environmental program, "which sold them and is using the $400,000 proceeds to fight global warming."

Goodwin made her decision after learning how the energy giant had covered up its awareness of the developing climate change crisis for decades. Until this revelation, "I thought the company was being foolish" in its insistence on sticking with oil and gas. "But we now know it was worse: it was being deceitful, in a way that is almost unimaginably heartless to future generations. . . . As the enormity of the effects of its lies becomes more evident, Exxon Mobil is positioned to supplant Big Tobacco as global Public Enemy No. 1."

Goodwin announced that she had "lost faith in Exxon's future value." And though she couldn't predict when environmental realities would finally have an impact of Exxon's stock price, she was nonetheless "glad that the recipients of my Exxon stock sold it immediately."

Her stand was followed in late March by an announcement from the Rockefeller Family Fund, a philanthropy established by the fourth-generation descendants. While following the lead of the Rockefeller Brothers

Fund (a product of the third generation) in divesting its assets from fossil fuels, the Family Fund went further when it came to ExxonMobil. "We would be remiss if we failed to focus on what we believe to be the morally reprehensible conduct on the part of ExxonMobil," the announcement said. The corporation had "worked since the 1980s to confuse the public about climate change's march, while simultaneously spending millions to fortify its own infrastructure against climate change's destructive consequences and track new exploration opportunities as the Arctic's ice receded. . . . Appropriate authorities will determine if the company violated any laws, but as a matter of good governance, we cannot be associated with a company exhibiting such apparent contempt for the public interest."

The Fund's $130 million in assets would be withdrawn from Exxon-Mobil, as well as all coal and tar-sands outfits, to be reinvested in other more socially responsible enterprises.

Soon after, at the May 2016 ExxonMobil annual meeting, shareholders for the first time passed a measure allowing investors who own three percent or more of company shares to nominate outsiders for the company's board. This might even open the door some day for a climate activist to become an ExxonMobil director.

Then, in the fall, the Rockefellers lay down the gauntlet with a dissection of Exxon's climate coverup in the *New York Review of Books* headlined "The Rockefeller Family Fund vs. Exxon." Its co-author was David Kaiser, a fifth-generation descendant, who said in an interview: "We think it [the story] will make clear to the public that the so-called debate over climate science has been a fake one, artificially manufactured, and a basically dishonest one from the beginning. . . . The Rockefeller Family Fund has exercised its freedom of speech in expressing our repugnance at ExxonMobil's behavior." The direct descendants of John D. Rockefeller and their clan numbers some 270, from whom Kaiser said he'd received "passionate support."

In July 2015, Christopher Lindstrom, 36-year-old grandson of David Rockefeller, was finishing lunch in a crowded organic restaurant just down the street from Manhattan's Rockefeller Center. That afternoon, he was scheduled to be the featured panelist at a workshop on bioenergy at the United Nations headquarters. The panel was part of the annual gathering of Nexus, an organization of young heirs from around the world who were committed to investing their family fortunes in environmentally sustainable ways.

"When the RBF signed onto the divestment movement, that had a pretty big effect," Lindstrom was saying, "much larger than the amount of money they were actually divesting. Because a lot of institutions were like, well, if they're doing it, why not us? The amount of money that institutions collectively are promising to take out of the fossil fuel industry is, I heard, up to $1.6 trillion. That's a lot of money. The question is, where is that money going to go? It's relatively meaningless if you put it into big pharma, or big agriculture. So my work right now is trying to expand the conversation toward shifting our world view in the way human beings relate to the earth and to each other—to contribute to a total transition."

Lindstrom came by his activism at a young age. "I inherited my mom's anger," he said, referring to banker David Rockefeller's daughter, Abby. "She was already very aware of the injustice of capitalism, and the ecological destruction caused by it. So I grew up having that conversation as a kid. I guess later on in my life, I became skeptical of the political focus of my parents' generation. My mom and stepfather and a lot of their peers were fixated on trying to solve things through the electoral system. But starting in my late teens, I got really disillusioned with that, seeing corruption even among the best candidates. The politicians who survived were those willing to compromise, and often by the time people rise to a level of high office, they've really sold their soul. I felt I wanted to make a big systemic shift through the avenue of entrepreneurialism and investing."

When Lindstrom read a college thesis written by his grandfather, it led to the opening of a dialogue between them. David Rockefeller had published his dissertation for the London School of Economics in 1941, titled *Unused Resources and Economic Waste*. "The book has zero ecological awareness, it's coming from a purely economic perspective," Lindstrom said. "Still, its subject was how much financial resources were wasted through our current system, that it's not efficient in distributing resources. So I talked to my grandfather about this, because he did demonstrate some awareness of that fundamental flaw."

Lindstrom went on: "Obviously my grandfather is the last iconic member of the Rockefeller dynasty. He has an identity with the system, to the degree that he was very involved with its creation. I've struggled my entire adult life with finger-pointing at an individual, because inevitably I come up against a group of people who want to cast all blame on a few faces. I would say, there's no one face to blame. And I told my grandfather about this.

"At this point, he just turned 100. I don't know if he's really going to wake up, I mean he's surprisingly lucid but he puts a large amount of resources into creating an environment of comfort and leisure, which feels to me is a kind of anesthetic to what's actually happening. I think that's the reason we're in this problem in the first place, that the people at the top use their resources to desensitize themselves to the pain of what the world is actually going through. Because right now everybody is stuck within the shell of a certain world-view. To a certain degree I am too, because although I might be thinking about systemic issues and structures, there's still an inner reality which is connected to scarcity and fear and self-worth and alienation, existential issues that are connected to the world we live in, the world of separation. But I'm hoping at some point to make a presentation to my family to help crack the egg."

The conversations and arguments over climate change and the future of the planet that are going on within the Rockefeller clan mirror those occurring in families across America. But they carry a striking poignance in a family whose name is inseparable from the fossil fuel age.

When Lindstrom arrived at the UN later that afternoon to address the Nexus organization, he found a full crowd waiting to hear his presentation. He began with his family background as a fifth-generation descendant of John D. Rockefeller, but also someone whose grandfather (on his father's side, the Lindstroms) invented the first composting toilet. He spoke of these as two opposing paradigms: the former based on an extractive economy that viewed the earth's biological resources as infinite, the latter a philosophy that recognizes the cycles of life and the need to integrate waste into what Lindstrom termed "the soil economy."

Lindstrom made clear that his ambitions went beyond those of most liberal heirs: "Going a few steps beyond just investing in do-gooder companies, how can we invest in companies that actually start moving resources into the transformation of the system itself? The first real necessity at this point is to compost the bullshit in our heads about the world we live in, that we're at the end of the line of a chain of living creatures and everything is here to benefit us. We have the technology to create renewable fuels that's been suppressed by the fossil fuel industry." Lindstrom paused momentarily before continuing. "Because there is an entrenched status quo, grandparents of people in this room, who have been benefiting from the standard of oil."

The roomful of next-gen inheritors didn't miss the double entendre. Lindstrom laughed, too. Energy, he well knew, was the bedrock of it all. Bioenergy, he also knew, had a bad reputation from a decade earlier, when corn was suddenly associated with fuel instead of food. No one, however, seemed to be looking at the true regenerative potential of bioenergy—taking waste and turning it into energy, while also producing a byproduct that's a source of fertilizer. In other words, an integrated bioenergy system that creates a carbon-negative energy economy, putting more carbon back into the ground than you take out.

At the UN that day, Lindstrom outlined an ambitious venture to turn a massive landfill in Kingston, Jamaica, into a waste-to-energy biomass facility that would power much of the country's grid. Eventually, the promised investors backed out, but that hasn't deterred the Rockefeller heir from his quest for a new energy economy. He's lately been discussing a cooperative venture to revive the North American hemp industry, starting in Colorado.

"It's a little-known fact that industrial hemp sequesters four times the carbon of planting trees," Lindstrom says. "You can also grow hemp in poor soil and enhance the fertility, in areas where other crops won't grow. And it's a hardy plant that in theory doesn't require any pesticides. So think about areas that are experiencing desertification and other side effects of climate change. Planting hemp could transform economically disadvantaged countries."

Lindstrom is again eyeing Jamaica, where a young investor he met at Nexus is interested in converting the sugar cane-based economy to growing hemp. Besides substantial potential revenue on the medical side, hemp can function as a highly nutritious feedstock and serve as a bioenergy infrastructure that includes storage, batteries, paper, textiles, bioplastics and more.

Lindstrom believes "that we've been brought to this crisis as a sort of evolutionary catalyst. Ultimately, I think that the shift is moving from a world rooted in separation and fear to a world rooted in the awareness of the connectivity of life."

At the 2016 shareholders meeting in the Dallas Symphony hall, Exxon-Mobil CEO Tillerson casually told the audience, "There is no alternative fuel known in the world that can replace the pervasiveness of fossil fuels." But his company has known for decades that's simply not true.

Joe Maceda, now in his early 60s, runs a New Jersey-based company called Gibbs Energy. And he's willing to talk about how today's energy equation might have been very different. In the early 1990s, Maceda had given a presentation to the South Coast Air Quality Management District in Los Angeles on an inexpensive form of fuel cell technology. As he left the auditorium, a short, bearded gentleman approached and invited him to lunch. His name was Patrick Grimes, and he worked under Rex Tillerson at Exxon Enterprises.

"That was a wild group," Maceda recalled. "It did everything. They'd invented the first word processor and the first fax machine. One of the projects that Exxon R&D hired Pat for was co-development with Alsace in France on flow batteries. Pat ended up being a one-man business unit reporting to Rex Tillerson, for 17 years."

Dr. Grimes, as Maceda remembered, had grown up "a farm boy in Iowa who used to cook over corn cobs, a common fuel on farms there." By the early 1960s, Grimes was a scientist at Allis-Chalmers, which had already introduced the first fuel cell-powered tractor. Grimes was the first to explore the possibility of recycling carbon dioxide to make fuel. In 1968 he'd patented an invention to capture or entrap atmospheric CO_2 and also described a method of sequestering CO_2 in ocean waters involving "the calcining of a material, such as limestone, dolomite, or carbonate to capture CO_2 from the air, for ultimate disposal in the ocean."

According to Maceda, "Grimes' experiments involved the electro-chemical reduction of carbonates, and he found an interesting range of hydrocarbon products—one might call this electrochemical fuel synthesis. . . . However, with discouraging oil prices, expanding oil reserves, and no concern about the atmosphere, management found his results uninteresting."

By the time Grimes got hired by Tillerson at Exxon, "they were getting out of everything other than what they do," Maceda said. "Which is how the system is set up. Not that they lie awake at night deciding how to fuck people, it's the natural result of the structure and the incentives. Forget R&D. When I first met Pat, the R&D guys at oil companies were startlingly better than the auto guys. That's long gone. They cut [R&D budgets] 90 percent and then another 90 percent. The smart guys all left, and left with the technology. Because oil didn't know what the hell to do with it. You can license anything from Exxon. They filed thousands of patents."

And let them languish. The Center for International Environmental Law (CIEL) revealed the depth of this deception in releasing a trove of previously unseen patent records in May 2016. These showed that Esso, one of ExxonMobil's precursors, had secured at least three fuel cell patents as early as 1963. Company scientists created early versions of the batteries now used to power electric vehicles like the Tesla. At the same time, according to CIEL president Carroll Muffett, Exxon executives "then turned to Congress and said you don't need to invest in electric vehicle research because the [company's] research is ongoing and it's robust."

By 1978, Harold Weinberg, a manager at Exxon Research, was advising the company that climate change "may be the kind of opportunity that we are looking for to have Exxon technology, management and leadership resources put into the context of a project aimed at benefiting mankind." Electric vehicle research was certainly one area that could have benefited from increased Exxon investment, thereby "benefiting mankind."

That same year, Exxon's labs came up with a novel device to help power electric motors for hybrid cars. In a glossy brochure, Exxon exulted that "the prototype has been engineered, tested, driven, proven." Work had begun four years earlier, fueled by concerns that oil might soon run out. Apparently US carmakers didn't express much interest, but Toyota did. In 1981, Exxon engineers dispatched a hybrid gas-electric Toyota Cressida automobile to Japan, with technology enabling an alternating current (AC) motor—which is less expensive, smaller, and more reliable than a DC motor.

As *Inside Climate News* first reported last October: "Despite that triumph, Exxon soon scaled back its investments in alternative energy, largely held by its venture capital unit, Exxon Enterprises Inc. (EEI). A drop in oil prices cut into budgets that financed EEI. And a rising generation of top management wanted Exxon to return to its core oil and gas business, rather than fashion itself into a comprehensive energy company. Exxon sold its licenses for the battery division's work and dismantled the electric drive team that built the hybrid Cressida. Sixteen years later, when Toyota launched the world's first mass-produced hybrid car, the Prius, it was equipped with an advanced AC motor."

It would be 1997 before Exxon announced a multimillion-dollar fuel cell-related research alliance with ARCO and a division of General Motors, Delphi Energy & Engine Management Systems. That same year, Daimler-Benz announced a $295 million investment in fuel cell technologies. Nearly two more decades passed before ExxonMobil, in May 2016, announced that for the past five years it had been engaged in joint research with the FuelCell Energy company on routing CO_2 from fossil fuel-burning power plants into fuel cells to produce electricity that would eliminate 90 percent of the CO_2 emissions.

By then, Pat Grimes had retired from Exxon, but he continued to file a series of fuel synthesis patent applications, using CO_2 and carbonates as feedstocks to produce hydrogen fuels. CO_2 vented to the atmosphere still contains significant energy, and how to capture it is well understood. "Most of this chemical heat is readily recoverable," according to Maceda. "The most advantageous process would be one, during fuel synthesis, that converts the thermal energy into chemical energy in the fuel product. As Dr. Grimes once summarized his approach, 'Use heat to make chemical bonds.'"

Recalling the pioneering man who once told him his job was "to sort of think about the future," Maceda shook his head. Maceda was convinced that his old mentor had found the solution to the climate change crisis, but his efforts had been stifled by his employer, Exxon. "I asked him one time, Pat, you've had the solution to the world's problems for more than 50 years, why the hell haven't you done it? He said, 'I need a job.' He had six kids. He drank and he died early, he was so frustrated. Nobody would listen to him."

Maceda paused to consider the wonders of the technology that Grimes tried to pioneer. "Even with natural gas, you could have 40 to 50 percent less of a carbon footprint without getting smart. With coal, I could have a zero carbon footprint because it's about sequestering of carbon. You can actually make the best biochar [charcoal used as a soil amendment] ever out of coal. If you wanted to reclaim deserts, buy a coal mine. Nothing in the world like it. That stuff holds water like nothing you've ever seen."

What stops researchers like Grimes and Maceda is not the limits of science and technology. "What we're up against has nothing to do with what is technically feasible, it's what is *politically* feasible," said Maceda.

"You're dealing with entrenched people who are living really well. Once the methane starts coming out of the permafrost, man . . ." Maceda paused. "But I wouldn't count on anybody in the finance or political world doing anything but what they're doing now. I mean, they're still denying climate change because they get paid to."

Chapter 4

The Fossil Fuel Companies' PR Maestro— and the Son Who Denounced Him

In December 2014, at a conference in Colorado Springs sponsored by the Western Energy Alliance, a roomful of fossil fuel executives gathered inside the Broadmoor Hotel and Resort. ExxonMobil had a representative on-hand, as did Halliburton and Devon Energy and many more companies involved in the extraction business. The speaker, public relations wizard Richard Berman, had flown in from Washington to discuss the latest tricks of the trade, as practiced by his firm, Berman and Company.

At age 72, the bald-headed, six-foot-four Berman still cut an imposing figure. And he did not mince words with his audience. "You can either win dirty or lose pretty," he told the energy execs. "Think of this as an endless war. And you have to budget for it." Berman didn't realize that his talk was being audio-recorded by one of the attendees, who anonymously provided a copy to the *New York Times*, explaining, "That you have to play dirty to win, it just left a bad taste in my mouth."

Using whatever underhanded tactics he could muster, Berman had long profited from the largesse of many in the room. He resided with his second wife in a $3 million, 8,800-square foot mansion in suburban McLean, Virginia, where he decided each morning whether to drive a Ferrari or a Bentley to his office in the nation's capital. There, some 30 employees occupied the entire eighth floor of a tall building on K Street. One of their main assignments consisted of how to defeat the environmental groups calling for stricter regulations and alternative energy development.

For the oil and coal representatives in Colorado Springs, Berman offered a primer on his methodology. Exploit emotions like fear, greed, and anger, he counseled—anything "to diminish the other side's ability to operate. . . . You could not get into people's heads and convince them to do something as easily as you could get into their hearts or into their gut to convince [them]. . . . Emotions drive people much better than intellectual epiphanies. If you want a video to go viral, have kids or animals," showing a video clip that used schoolchildren.

"Use humor to minimize or marginalize the people on the other side," Berman continued. His company's latest front group was called "Big Green Radicals." Among its ploys were sending a "poison valentine" to the fossil fuel divestment campaign, claiming that environmental groups receive funding from Russian oil interests tied to President Putin, and putting up billboards on the Pennsylvania Turnpike attacking Robert Redford, Lady Gaga, and other celebrities for opposing fracking.

"There is nothing the public likes more than tearing down celebrities and playing up the hypocrisy angle," asserted Berman's corporate vice-president, Jack Hubbard, who accompanied him to the meeting. For example, a billboard with a picture of Redford was emblazoned: "Demands green living. Flies on private jets." Hubbard also told the energy executives gathered at the Broadmoor Hotel that he'd been doing research into Sierra Club and NRDC board members, looking for embarrassing tidbits on their personal lives.

Summing up how his game worked, Berman reassured the energy titans that they would not be soiled by his dirty work. "People always ask me one question all the time: 'How do I know that I won't be found out as a supporter of what you're doing?' We run all this stuff through non-profit organizations that are insulated from having to disclose donors. There is total anonymity. People don't know who supports us." This, undoubtedly, was welcome news to those in the audience, many of whose companies had signed six-figure checks to Berman, the master of disinformation.

Rick Berman—who was dubbed "Dr. Evil" in a 2007 *60 Minutes* report for his similar appearance (and apparent morality) to the *Austin Powers* movie villain—was raised in New York City. After gaining a law degree from William and Mary Law School, Berman became director of labor law with the US Chamber of Commerce by age 30, before moving on after two years to represent the restaurant industry as a lead lobbyist. He founded his own consulting firm in 1986, initially helping restaurant chains fight against minimum-wage campaigns, block food safety regulations, and curtail efforts to limit smoking in their facilities. During the 1990s, Berman formed Beverage Retailers Against Drunk Driving (BRADD) to combat Mothers Against Drunk Driving (MADD), with his group arguing for "tolerance of social drinking."

Also in the 90s, Berman began creating a number of nonprofit charitable groups, all run out of the same Washington address. These could take large tax-deductible sums from donors whose identities remained secret. The first was the Employment Policies Institute, followed by the Environmental Policy Alliance—and ultimately at least 40 more linked organizations. Their board members were mostly current or former employees of Berman's firm, or consultants for the various industries represented. A web site proclaimed: "Berman and Company isn't your average PR firm. We don't just change the debate. If necessary, we start the debate."

Berman pushed hard to help the tobacco industry win the debate. His Guest Choice Network (later renamed the Center for Consumer Freedom) was incorporated with a $600,000 donation from Phillip Morris, which later contributed another $2.3 million.

"Berman's a hired gun, he goes where the money is," says John Dunbar of the Center for Public Integrity. "Nothing he does surprises me. . . . Berman's strategy is to go to the masses and persuade them to pressure office-holders. Since he's not contacting lawmakers, he doesn't have to register as a lobbyist and say who he's working for or how much he's getting paid."

In recent years, money has been pouring into Berman's coffers to counter the environmental movement. Besides Big Green Radicals, he runs a media operation called EPA Facts that targets the federal environmental agency. In a full-page ad that ran on the *Politico* web site in 2014, EPA Facts asked, "What would you call a radical organization that threatens to shut down 25% of our electrical grid?" The words anarchist, terrorist and militia were X'd out, with Obama's EPA the "correct" answer.

LEED Exposed, another Berman web site, attacks certified green buildings as a waste of money because they are "not necessarily more energy or water efficient than other buildings." Berman's Green Decoys takes on sportsmen's groups for backing federal attempts to reduce air pollution and carbon emissions. "Their claims are so far-fetched and their tool more than anything else is misinformation," declared Chris Hunt, Trout Unlimited's director of communications.

Then there are Berman's attacks on energy policy—any policy except one favoring fossil fuels. Many of these critiques of renewable energy come in the respectable packaging provided by industry-funded research institutes. Around the corner from the Boston Statehouse, the Suffolk University economics department hosts a conservative think tank known as the Beacon Hill Institute. One of the few conservative research institutes in the liberal bastion of Boston, BHI was founded in 1991 by businessman and Republican politician Ray Shamie, who stated its mission was "grounded in the principles of limited government, fiscal responsibility and free markets." Today, many of BHI's sham "studies" are funded by Berman.

BHI's executive director is Peter Tuerck, who until 2013 chaired the economics department at Suffolk University and remains a professor there. He also appears on the roster of experts at the Heartland Institute, where he's spoken at conferences on the economics of climate change. At one such conference, in 2009, Tuerck explained, "I have found it necessary to go around the country pointing out that claims about green jobs are all phony."

In 2014, BHI took in about a million dollars toward research, including funds from a Berman front group called Employment Policies Institute, the Searle Freedom Trust (a conservative foundation that gives to numerous right-wing think tanks), the Koch Foundation (over $750,000 in donations altogether), and the Coors family's Castle Rock Foundation.

In 2011, BHI began honing in on attacking renewable energy standards in various states. Here's a classic example of how its strategy works:

1. BHI does a study on, for example, Maine's new law requiring power companies to obtain part of their electricity from renewable sources; predictably the think tank's analysis of the state law is very discouraging, estimating that it's going to cost ratepayers $145 million and almost a thousand jobs within five years.

2. The report is republished by the Maine Heritage Policy Center, a conservative advocacy group that, like BHI, belongs to the State Policy Network, an umbrella group for corporate-sponsored think tanks seeking to influence what happens in state legislatures. (The Koch brothers are among the State Policy Network's principal funders.)

3. The BHI report is immediately touted by Maine's Republican Governor, Paul LePage, a staunch opponent of energy-efficiency programs. In 2016, the governor comes out against a legislative plan to increase solar development in Maine twelve-fold over the next five years, a program that its proponents argue will create new jobs and cut consumer energy costs. Instead, LePage pushes

for incentives encouraging the expansion of a natural gas pipeline.

4. The BHI study is lauded by the American Legislative Exchange
Council (ALEC, also Koch-funded) and the Heartland Institute,
which have written their own Big Energy-friendly "model legislation" known as the Electricity Freedom Act. Meanwhile, the
State Policy Network makes sure that BHI reports and ALEC's
model energy bill get widely distributed in state capitals where
renewable energy initiatives are being considered.

BHI's anti-alternative energy report has popped up all over the country, including New Jersey, where the study was released by the Koch
brothers' nonprofit Americans for Prosperity, in response to the state's
plan to develop 1,100 megawatts of wind energy off its coastline. While
stating that global warming remained a matter of "considerable debate,"
the BHI study concluded that the wind farm would "produce net economic costs, raise electricity prices, and dampen economic activity."
Nothing was said about some 5,000 construction and maintenance jobs
projected to result from the wind project, or the benefits involved in
reducing New Jersey's need to import electricity from other states. On
it goes—BHI makes its influence felt in state capitals from North Carolina, to Ohio, to Wisconsin.

Kalin Jordan, a Suffolk University graduate in political science (class
of 2009), stumbled upon BHI and its funding sources several years later:
"What I found is, groups in other states work with BHI to write the
report on renewable energy, but it has the credible name of Suffolk University attached to it. Nobody knew that it was Koch money or Berman,
so no one questioned it." Jordan, soon realizing this was an issue that
went well beyond her alma mater, formed a new organization called
UnKoch My Campus. She also managed to convince Suffolk's president
to release a list of BHI's funders. The university administration even
stepped in to block a BHI grant from the Searle Freedom Trust in 2013,
which was to fund a propaganda campaign against the Regional Green-

house Gas Initiative, a joint effort by Northeastern states to reduce carbon pollution.

This proved to be a temporary setback for BHI. In June 2014—shortly after the EPA proposed its Clean Power Plan, a landmark effort to lower the carbon emitted by power plants—BHI used a $41,500 grant from a Berman front group to mount a series of opposition studies in 16 different states. At the same time, Berman secretly routed funding to create five new front groups.

BHI's first "academic" study appeared in January 2015, a 29-page "national report" claiming that "the Obama administration has unveiled an unprecedented scope of regulation," including "CO_2 emission limits on new and existing electricity power plants and new lower limits on mercury emissions." Predictably, the BHI report went on to warn of dire consequences for the energy industry and consumers.

The national release was followed by statewide reports using the same BHI data on the costs of the EPA rules—in New Mexico (where the report was released by the Rio Grande Foundation); Wisconsin (the MacIver Institute); North Carolina (the Civitas Institute); Iowa (the Iowa Public Interest Institute); Louisiana (the Pelican Institute for Public Policy); South Carolina (the Palmetto Promise Institute), and Virginia (the Thomas Jefferson Institute for Public Policy). All of these BHI partners were right-wing think tanks. And BHI's reports, not surprisingly, were a mirror image of ALEC's efforts to introduce "model bills" in legislatures of the same states.

When compared to the EPA's analysis, the BHI studies inflate the costs of the clean energy measures by two times and minimize the benefits by a factor of ten. Dr. Frank Ackerman is a Harvard Ph.D. and now a principal with Synapse Energy Economics, Inc., who has written extensively on the economics of climate change. Asked for comment on the BHI studies, Ackerman responded: "They're almost too easy a target in terms of intellectual content. The approach to economic modeling which BHI uses, known as STAMP [State Tax Analysis Modeling Program], does not appear in the academic literature and is not used by much of anybody else. STAMP has never seen a government program that it liked or a tax cut that it disliked."

According to Gabe Elsner, executive director of the Energy & Policy institute, "The attack on the EPA seems to be coordinated among BHI, State Policy Network groups, and the American Legislative Exchange Council, and I think Berman is at the heart of it."

In the end, BHI's increasingly controversial public profile took a toll on its relationship with Suffolk University. In December 2015, BHI announced it would be leaving by the end of the following year. "I couldn't raise money under the guidelines that were being issued," said executive director Tuerck. So he asked the school for an "amicable divorce." The likelihood is that BHI will continue operating as an independent.

Rick Berman, too, has become increasingly notorious. His studies and "information" campaigns have been exposed as brazen corporate propaganda, and his own ethics have come under fire by watchdog groups like Citizens for Responsibility and Ethics, which filed a complaint with the IRS charging him with funneling millions from his Big Energy advocacy campaigns into his own pockets. While federal tax laws don't allow the creation and operation of nonprofits to benefit private interest and individuals, at least one of Berman's established "charities"—the Center for Consumer Freedom—paid 92 percent of its $1.4 million in donations in 2011 to Berman and Company for "staff[ing] and operat[ing] the day-to-day activities."

In 2016, Berman turned his media guns against Republican presidential contender Donald Trump, even writing an op-ed attack against the GOP candidate for the conservative *Washington Times* under his own byline, in which he called Trump "the embodiment of a decaying political culture that prizes celebrity over leadership." The op-ed caught the eye of the OpenSecrets Blog, which revealed in March that a Berman front group called Enterprise Freedom Action Committee had by then spent $315,000 on anti-Trump Google and Facebook ads. Where the funding for the ad campaign came from was unknown, since the committee is a "dark money" nonprofit that needn't make such disclosures.

But the Koch brothers were among the wealthy clients of Berman who had decided Trump was then too much of a wild card to be worthy of their support, despite his boisterous calls for more drilling, mining and fracking. Hillary Clinton, whose campaign was awash in oil money, was viewed as a perfectly acceptable alternative in energy industry boardrooms across America.

Men like Rick Berman easily cross party boundaries; they can win with Republicans or Democrats. The PR master exudes the confidence of a man who has long been used to winning. But there is one chink in Berman's armor. It's his own son.

David Berman is his father's "inconvenient truth." Unlike many children of fossil fuel families, who either shy away from public comment or more often follow their father's footsteps, Berman's son has publicly denounced him.

For 20 years, David Berman was the lead singer for a popular indie rock band called the Silver Jews. He also penned an acclaimed book of poetry, *Actual Air*, and contributed poems to a book about the artist Friedrich Kunath titled *You Owe Me A Feeling*.

Feeling, it seems, might have been sorely lacking in the household where the young Berman grew up. Born in 1967 in Williamsburg, Virginia, he attended high school in Texas before graduating from the University of Virginia. There he began writing and performing songs, eventually moving to New York and recording six albums with the Silver Jews on the Drag City label.

In 2003, Berman underwent a severe period of depression and substance abuse, even attempting suicide. He later described this period as "an incredible blessing" that led him to deeper involvement with Judaism. Two years later, the reunited band made its first tour.

Then, on January 22, 2009, the singer posted a note on the music group's online message board, announcing that he was retiring from making music. The reason? "Now that the Joos [sic] are over I can tell you my greatest secret," Berman wrote. "Worse than suicide, worse than crack addiction: My father. You might be surprised to know he is famous, for terrible reasons. My father is a despicable man. My father is a sort of human molestor. An exploiter. A scoundrel. A noted historical mother ****** son of a bitch (sorry grandma). You can read about him here: www.bermanexposed.org."

Berman, 42 at the time, continued: "A couple of years ago I demanded he stop his work. Close down his company or I would sever our relationship. He refused. He has just gotten worse. . . . More powerful. We've been 'estranged' for over three years."

Even as a child, David said, they were opposites. After college, he joined the Teamsters, knowing that his dad was out to destroy the unions—not to mention targeting "animal lovers, ecologists, class action attorneys, scientists, dieticians, doctors, teachers." As his anguish over his father's ways grew, David immersed himself in the study of Judaism but still "could find no relief." The Silver Jews' creative contribution to the world was too small of a mitzvah up against "all the harm he has caused," David continued in his soul-baring message to his fans. "I've always hid this terrible shame from you, the fan. . . . In a way I am the son of a demon come to make good the damage. Previously I thought, through songs and poems and drawings I could find and build a refuge away from his world. But there is the matter of Justice"—which for David was like a "burning pain. . . . There needs to be something more. I'll see what that might be."

A week later, the Silver Jews performed for the last time before several hundred fans inside Tennessee's Cumberland Caverns, located 333 feet underground. Now living with his wife in Nashville, Berman declined an interview request for this book, saying in an e-mail: "I've been writing, or should I say writhing about my father and what he represents for six years now. Minus two years of total surrender distributed over six month periods, it's safe to say that my interest in living out the second half of my life is inordinately contingent on me solving my labyrinthine dilemma."

The "labyrinthine dilemma" posed by Richard Berman and his colleagues is one shared by many in David's generation whose fathers bear various degrees of responsibility for the climate havoc now imperiling life on the planet. As these sons and daughters grow older themselves and envision the rest of their lives and the lives of their own children, a heavy weight hangs on some of them, perhaps an extra burden to do good in the time that remains. "I look into the mirror, but it's cracked," writes David in "The Broken Mirror." "My life is almost over; that's my fact."

Chapter 5

The Man Who Pioneered Fracking, and a Granddaughter's Quest

For the descendants of fossil fuel families, talking about their lives and their heritage doesn't come easily. Many have simply followed their fathers' lucrative but stained paths, the sons of Lee Raymond and Rex Tillerson among them. Even those who harbor dark anxieties about their family legacies generally prefer to keep these feelings to themselves. Very few we sought out for this book were willing to go on-the-record about their families.

Anna Getty is one of those who couldn't keep silent. The adopted great-granddaughter of J. Paul Getty, the oil tycoon once crowned the richest man in the world, has grown increasingly appalled by the behavior of big oil companies and other corporate giants. "It's shocking and atrocious, and bad karma if you want to get into the spiritual sense," she told the press, "All I can think is that [these corporate executives] just don't care. They don't care what is going to be left in 50 years, because they're all going to be dead."

Now in her forties, Getty has long had a passion for green living and a holistic lifestyle. She's served on the boards of numerous environmental organizations, including the Environmental Working Group and Global Green. She quietly helps underwrite documentary films like *Fuel*, which promote renewable energy and biofuels as replacements for fossil fuels. Her second home, in Ojai, California, is off-the-grid and hosts an organic farm.

Getty is not the only oil heiress who has devoted herself to saving the planet. Based in western Massachusetts, Jenny Ladd, a Standard Oil heiress (her great-grandfather Charles Pratt became a partner of John D. Rockefeller in the 1870s), pursues a variety of philanthropic green activities, contributing to groups like Greenpeace, Bill McKibben's 350. org, and an indigenous organization fighting the expansion of Canadian tar sands mining operations.

Leah Hunt-Hendrix is the granddaughter of Texas oil tycoon and notorious right-wing funder, H.L. Hunt. Now living in Brooklyn, she describes herself as a radical philanthropist and was deeply involved in the Occupy movement and the New Economy Coalition. She wrote in *The Nation* in 2014, "We can start by shifting the endowments of foundations and universities away from investment in fossil fuels and private prisons and into a renewable economy owned and operated by communities, with members accountable to each other and to future generations."

Among the most interesting heiresses to a fossil fuel-related fortune is Katharine Lorenz. Since 2011, Lorenz has overseen the Cynthia and George Mitchell Foundation, whose primary goal is to ensure that the natural gas shale drilling pioneered by her grandfather doesn't exacerbate climate change and other environmental problems.

Lorenz's Austin, Texas-based foundation has funded major studies on the impact of methane emissions caused by hydraulic fracturing ("fracking," as it's better known), and on how Texas can best meet the EPA's Clean Power Plan. The foundation has also tried to find common ground between environmentalists and conservatives, hosting a gathering of those on the right who called for an end to fossil fuel subsidies and a tax on carbon pollution.

Lorenz's biggest inspiration remains her grandfather, George Mitchell, who died in 2013 at the age of 94. The son of a Greek immigrant who ran a dry cleaning business in Galveston, Mitchell graduated at the top of his class at Texas A&M, served as an Army engineer in World War II, and then started "wildcatting" during the postwar natural gas boom. At the time, Mitchell's wells helped coal-burning cities like Chicago and Pittsburgh move to a cleaner source of heating fuel. By the late 1970s, he was a wealthy man, and decided to experiment with a new method that combined two technologies—hydraulic fracturing and horizontal drilling, injecting a mix of water, sand and chemicals under high pressure to release the gas trapped in hard-rock shale formations.

"So many of his peers thought he was nuts," his granddaughter recalls. It took 16 years, and a $250 million investment, before Mitchell's innovative approach paid off. In the late 1990s, when other companies began successfully employing his method, the environmental problems since associated with fracking had yet to surface. Mitchell's brainstorm seemed not only a cleaner, less carbon-emitting fuel source than oil, but a path to greater energy independence.

Ironically, perhaps, Mitchell's other passion was sustainability. It began in the 1960s, when he heard Buckminster Fuller speak and visited the Watts district of Los Angeles not long after the rioting there. "He came back from that saying we're doing it all wrong and have to find a better way to make communities work," says Lorenz. "Let's work toward a walkable, livable integrated society."

Mitchell consulted architects involved in creating intentional communities. Then he purchased The Woodlands, 28,000 acres of timberland about 30 miles north of his modest home in Houston. After securing a $50 million loan guarantee from the federal Housing & Urban Development agency, the innovative community evolved into a patchwork of neighborhood villages nestled between green spaces, including neighborhood parks, bike paths and hiking trails. "In the early days, if he drove down the street and saw someone mowing grass on the median, he'd pull over and scream at them," Lorenz says.

In 2011, Mitchell sold The Woodlands, which now has a population of 100,000 and has since become the home of a new corporate headquarters for ExxonMobil. However, another 5,600 acres bought by Mitchell for a family ranch have been restored to pre-settlement conditions. While surrounding ranches were clear-cut, Mitchell preserved the forest as a nesting habitat for the formerly endangered red-cockaded woodpecker. The Cook's Branch Conservancy, as it's now called, has become an outdoor laboratory for research scientists, overseen by another of the Mitchell grand-daughters. In 2012, it received the Leopold Conservation Award for habitat management and wildlife conservation on private land.

For Lorenz, spending considerable time with her grandfather in the last years of his life proved the turning point in her own education. "I'm one of 25 grandchildren, so growing up I'd see him a couple times a year, generally in the context of a big family gathering," she said. "He was very quiet, very shy, not someone I really got to know. Not until I was an adult did I create a stronger bond with him."

It started when her grandmother became ill with Alzheimer's, and Lorenz's mother often went to visit her at the Mitchell home. Lorenz, who graduated from Davidson College with degrees in economics and Spanish, was living then in Oaxaca, Mexico, where she co-founded a non-profit working to stem malnutrition among the local children. Her work in Mexico, between 2003 and 2008, made her aware of the link between the changing climate and subsistence farming.

"I'd founded an organization focused on a grain called amaranth, which is high in protein as well as being local, with large implications for health and well-being. The people planted when they always had, and it didn't rain. The next year, all the fields flooded. Three years in a row, this went on. The farmers would say, the climate has never been like this. If the crops fail consistently, they're going to starve. So I'd be going to funders saying, 'Oh, sorry, yet again.' I was seeing a microcosm of what's recently been happening in Syria, where the drought is behind a lot of the migration to Europe."

Her grandfather had always been more interested in "big picture science, how do you change the systems. But I think he was kind of proud that I went and did something so crazy, working in rural villages." When she returned to the U.S, she began coming to dinner at Mitchell's home several times a week, where they talked about their mutual interests. "He was fascinated by land use, species extinctions, oceans and water quality, all of it, and the importance of being in harmony with nature. And he certainly came to understand, when considering the sustainability of the planet, that climate change was going to be the biggest impact. I remember a few occasions when he said, 'That's the biggest threat, if we don't get that one right.'"

As far back as the 1970s, Mitchell had led conferences bringing together CEOs to discuss the role of business in addressing environmental challenges. He would pass out books on various environmental subjects. "He was willing to stand up and say what he believed, even in the face of people who didn't. My guess is, a lot of them thought he was completely nuts."

While Mitchell was close to former president Jimmy Carter, he didn't have kind words for the Bush-Cheney administration. Asked whether he ever expressed any doubt or regret about having pioneered fracking, Lorenz replied: "He was very vocal about the regulators needing to be more effective. I don't think he believed the whole idea of extracting gas or oil from shale was inherently wrong, rather that technical issues needed to be addressed. The independents, he'd say, will cut corners and screw it up, hurt people and the environment and basically ruin the whole thing."

After several years as deputy director at the Institute for Philanthropy, in 2011 Lorenz assumed leadership of the Cynthia and George Mitchell Foundation. Her grandfather was in his last two years of life, and their conversations honed in on what best to do with family resources that will approach $900 million by the time the estate is settled in 2018. While Mitchell's emphasis had been on funding large institutions working on sustainability, he and Lorenz agreed that different tools were now needed—"more grassroots organizations, more policy and advocacy work," in her words.

"The first grant-making program we did as a family focused on clean energy, as well as moving toward more efficient use of water because of the drought, and the driver behind this was really climate change," she said. "We also launched a program looking at shale sustainability issues, specifically around extraction. At the time, there was still very little information about methane and emissions."

Burning natural gas releases half as much carbon dioxide as coal, causing President Obama to tout fracking in 2012 as a clean fuel source, as well as a big job creator. But methane, the key component in natural gas, can leak invisibly into the atmosphere and is 25 times more potent a greenhouse gas than CO_2. Lorenz's foundation funded a study in Texas that sought to understand why leakage rates varied so much and how to address the issue.

The study, published in the prestigious journal *Science*, determined that "it's big [accidental] leaks, rather than a systemic problem," with the technology, according to Lorenz, "like someone leaving a valve open or not shutting something down. Human errors create the biggest problems. Another factor is how methane is measured, fly-over versus on the ground—being able to decipher methane coming from pipes and extraction operations, versus coming off of wetlands or a garbage dump or feedlots."

This led the family foundation, in 2015, to break with other oil-and-gas extractors who claimed that proposed EPA rules on methane would hike their costs and stifle the boom in domestic energy development. While the industry had made progress in cutting methane emissions through monitoring and leak-seeking infrared cameras, the Mitchell Foundation asserted that voluntary programs can only do so much and that stronger federal regulation is crucial because "the number of oil and gas companies that aggressively control their methane emissions must increase." This marked a shift from the foundation's earlier support of state-by-state efforts rather than federal mandates.

The disaster at the Aliso Canyon/Porter Ranch natural gas facility near Los Angeles, where a massive leak spewed gas for four months before it was plugged in February 2016, highlighted the urgency of regulating methane. Another study in *Science* determined that Los Angeles' methane output had doubled due to this leak alone, with some 60 metric tons released every hour. Obama has called for cutting methane emissions over the coming decade by as much as 45 percent from levels in 2012. The EPA is expected to finalize a rule on regulating methane for existing oil and gas wells in the near future.

Lorenz's foundation has also strongly urged the state of Texas to get on board with the EPA's Clean Power Plan, aimed at cutting greenhouse emissions significantly, which has provoked stiff opposition from energy companies and their front groups. "How do we get Texas to not only comply, but make adjustments and really excel?" Lorenz asks. "Part of our work is making sure that natural gas is just a bridge fuel, not an end in itself. We very clearly need to move toward a renewable future. We can't really get there *without* gas, but the transition has to happen as fast as possible."

Lorenz says that Texas is not a hopeless case, despite the dominance of the oil industry, pointing to the growing importance of wind energy in the state's grid. In fact, it was the second state after Iowa to implement a renewable energy standard, requiring a certain amount of electricity to come from such sources. Back in 2008, Texas invested millions in high-voltage power lines linking big cities to wind-filled areas in West Texas. Farmers were allowed to lease land to wind developers, bringing them another source of revenue. Nine percent of the state's electricity came from wind in 2014. Today, Texas leads all states with nearly 18,000 megawatts of installed wind power capacity, nearly triple that of its nearest rival, Iowa.

According to *Scientific American* magazine, in December 2015, a low-pressure weather system moving across the Texas panhandle brought sustained wind speeds between 20 and 30 miles per hour— enabling the state to set a new production record, with wind generating

45 percent of its electricity for many consecutive hours. The brownouts predicted by naysayers when wind is relied on as a primary power source didn't become reality. The magazine noted: "Texas was able to balance the intermittent wind because it has a lot of natural gas power plants, which can adjust their power output more quickly than coal-fired power plants. Considering this fact, it seems like a happy coincidence that market forces are transitioning the US electricity system toward a mix of renewable energy and natural gas."

This, too, is something that Lorenz's foundation has put money behind—studies showing that wind and natural gas can cooperate rather than compete, en route to a fully renewable future. To encourage this collaboration, the Mitchell Foundation brought together the key Texas players during a state visit by EPA administrator Gina McCarthy.

In 2014, Lorenz's foundation made another effort to expand the renewable energy coalition, sponsoring a symposium in Austin called "Getting Energy Right" which brought together a range of conservatives from across the country. Among the panel members was former six-term Congressman Bob Ingliss, a South Carolina Republican, who told the audience, "What we're looking for are energy optimists and climate realists." Ingliss proposed establishing a level playing field for all categories of energy producers, stripping the sizable advantages enjoyed by the fossil-fuel industries. We should "end all the subsidies [for oil, gas and nuclear], attach all the costs to all the fuels, and watch the free enterprise system sort all this out."

Debbie Dooley, a preacher's daughter who has been involved in conservative politics for four decades, also addressed the symposium. She was one of the 22 co-founders of the national Tea Party movement in 2009, and remains on its board of directors. But she has also fought successfully to make solar power mainstream in her native Georgia. "My message was: free market choice, competition, and empowering consumers," she said. In that very red state, Republicans control the legislature but have passed a Power Purchase Agreement bill allowing homeowners to install rooftop solar panels with no upfront costs and permit third-party leasing of solar.

"I have a grandson, Aiden, who is seven," Dooley said. "He'll know I fought for energy choice and freedom. He'll know I fought for a clean environment for him. I see it as my legacy to him. When you have a grandchild, you really start thinking more and more about the future."

Chapter 6

Harold Hamm and the Frack-Turing of Oklahoma

By November 2013, Austin Holland knew there was a big problem. For the past several years, he'd been the state seismologist with the Oklahoma Geological Survey (OGS), working out of a small basement office at the University of Oklahoma's College of Earth and Energy. Until recently, he'd been basically a custodian of geological records—cataloguing in logbooks the location and depth of a small number of earthquakes.

But to Holland's initial surprise, a spate of these tremors were suddenly plaguing the half-million largely rural residents of the central and north-central sectors of the state. Oklahoma had gone from experiencing a yearly handful of magnitude-three or higher quakes over the previous 30 years to more than a hundred quakes in 2013 alone. One of these, late in 2011, had registered 5.6 on the Richter scale in the small town of Prague—the largest in the state's history and cause of substantial property damage.

As he studied the disturbing pattern, Holland became convinced there was a direct link between the dramatic increase in earthquakes and the disposal of tons of wastewater from oil-and-gas fracking, pumped down into wells along a large network of faults that lay beneath the red-dirt prairie. The disposal method was aimed at avoiding contamination of clean water close to the surface, a growing problem as fracking sharply increased in Oklahoma, producing as many as ten barrels or more of the salty wastewater for every barrel of oil.

Holland was concerned enough to sign his name in 2013 to a joint statement with the US Geological Survey (USGS) suggesting that the state's "induced seismicity" was traceable to the fracking boom. Shortly after, the scientist was asked to "have coffee" with University of Oklahoma president and former US Senator David Boren and Oklahoma billionaire, university donor and fracker extraordinaire Harold Hamm. When Holland told a colleague about the coffee invitation from two of the state's most powerful men, he responded, "Gosh, I guess that's better than having Kool-Aid with them."

Since stepping down as a US senator to take the university position in 1994—paving the way for the election of notorious climate change-denier Jim Inhofe—Boren had raised a fortune for the school from men like Hamm. Boren was also appointed to the board of oil giant ConocoPhillips as a paid director for ten years, and then in 2009 to a similar position with Hamm's company—Continental Resources, the biggest independent fossil fuel company in the US Boren was richly rewarded for his corporate service. In 2014, he would receive- over $400,000 from his status as a Continental board member—more than his annual base salary as university president.

The coffee klatch in the president's office was, Holland said later, "just a little bit intimidating." Boren denied the meeting was an effort to pressure the state seismologist, later describing it as "purely informational," adding, "Mr. Hamm is a very reputable producer and wanted to know if Mr. Holland had found any information which might be helpful to producers in adopting best practices that would help prevent any possible connection between drilling and seismic events. In addition, he wanted to make sure that the Survey (OGS) had the benefit of research by Continental geologists."

But Holland knew that the oil industry had a long record of leaning on the Oklahoma Geological Survey. In February 2014, around the time of a class-action lawsuit initiated against two oil-and-gas companies by a woman whose home had been badly damaged in the Prague earthquake, the OGS issued a position paper asserting that the "majority, but not all, of the recent earthquakes appear to be the result of natural stresses." This concurred with the assessment of Continental Resources: that it was simply an unusually active period for earthquakes around the world.

Holland's boss at the time was Larry Grillot, dean of the College of Earth and Energy at the University of Oklahoma, as well as a generously paid board member of Pioneer Natural Resources, which happened to own coalbed methane wells in an earthquake-prone area of Colorado. Holland stayed publicly mum, at least for a time. Bob Jackman, a Tulsa-based petroleum geologist who'd grown concerned about the quakes, ran into the seismologist twice at meetings. "He gave a talk on the earthquakes at an oil-and-gas summit," Jackman recalled in an interview. "Damndest presentation I ever heard, because he'd show all this evidence pointing to fracking and disposal wells, and then stop before drawing the obvious conclusion.

"So afterwards I saw him in the hall and said, 'Austin, you're holding back.' He said, 'Bob, you don't understand. Harold Hamm and others will not allow me to say the obvious.' I said, 'Really.' Holland then went into this spiel: 'Our university needs these big donors, so there are certain areas where I used to say some things, but they won't let me.' He brought up Hamm's name twice. I realized an enormous cover-up was going on. Like the good whistleblower I am, I pulled out my little notebook and wrote down the whole conversation, then wrote an article that got published in the *Oklahoma Observer* [an op-ed monthly]. The headline was, 'Hamm-Made Earthquakes.'"

Six months later, as earthquake activity in Oklahoma jumped substantially again (567 quakes measured at 3.0 or more in 2014), Holland could no longer remain silent. He led a team of OGS scientists that sounded an alarm, releasing a public statement in April 2015 that the recent flurry of tremors was likely human-caused and tied to "injection/

disposal of water associated with oil and gas production." Not surprisingly, the statement produced tremors of its own. Dean Grillot promptly sent an e-mail to two colleagues, saying "Mr. Hamm is very upset at some of the earthquake reporting to the point that he would like to see selected OGS staff dismissed."

Seeing the writing on the wall, Holland left the University of Oklahoma, accepting a position with the USGS in New Mexico that July. He spoke out publicly for the first time in an interview with *Al-Jazeera America* for a documentary called "Fault Lines," recalling having received instructions from Hamm to "watch how you say things." He also described the OGS's official position on the earthquakes in 2014 as an effort "to slow things down, to give the deniers more time to deny." He called Dean Grillot's board membership with a fossil fuel company "probably a clear conflict of interest."

Meanwhile, despite the upheaval at the University of Oklahoma, President Boren continued to insist that nothing was amiss. "No researcher at the OGS . . . has ever received pressure from the university to change their research or slow their research," he contended. Boren conveniently overlooked the fact that while the OGS team based on the campus was formulating its industry-friendly position on earthquakes, Grillot was involved in a fundraising pitch aimed at getting a $25 million contribution from Harold Hamm for a new energy facility. (Plans for the facility were eventually dropped, reportedly due to the fall of oil and gas prices as well as a decline in the number of petroleum engineering students.) Boren acknowledged it was true that "most" of the $2.4 billion the university had raised in the course of his two decades as its president "has come from the energy sector." However, he maintained, "No university should ever take a donation where there is a quid pro quo that benefits the donor."

Austin Holland's position at the university, as of this writing, remains unfilled.

Harold Hamm has been called "the last American wildcatter." Moriah Stephenson, who's working on a master's thesis about the state's oil industry at the University of Oklahoma, finds it "interesting how Hamm embodies a narrative of Oklahoma identity and how that creates repression in the state. It's a particular rhetoric of the pioneering individual—a working class, low-income white male from a rural community who worked his way from bottom to top. As the richest man in the state, he's often used as this token symbol of what it means to be Oklahoman, so people justify his actions."

Despite the impressive energy fortunes that have been made here, Oklahoma remains one of the nation's most impoverished states. Once Indian Territory and considered largely useless land by white settlers, the accepted mythology is that "discovering oil and gas saved us from ourselves." But it's a false scenario, as Stephenson sees it. She views the fracking boom as a new Oklahoma land rush, the massive colonization of the state in 1889 by white families looking for cheap land, which displaced numerous Native American tribes. Only a few, like Harold Hamm, have become extravagantly wealthy, while the rest of the population must cope with the disruptions caused by the fossil-fuel and uranium industries.

Jesse Coleman, a Greenpeace-USA researcher and longtime Hamm-watcher, says: "Hamm is a very interesting character, because he comes from outside of the traditional executive class. With his brash cowboy attitude, he's not afraid to blow really hard about things, like attacking the US for its oil export and import policy. For someone who a lot of people might consider not sophisticated in his political viewpoint, he's really central to a number of important orbits of power. And Hamm has been extremely effective at creating a friendly legislative and regulatory environment for his company and interests. Taking his home base as a microcosm, nothing seems to happen at Oklahoma University without his express approval."

Hamm grew up the youngest of 13 children in a sharecropper family, with no indoor plumbing or electricity and sometimes no shoes. At six, he began skipping school and joined his father picking cotton. At 17, he left home and found a job pumping gas and fixing flats at a truck stop.

He also started privately studying geology. "It just grabbed my imagination," Hamm said years later, "that anybody could find this hidden, ancient wealth and it was yours."

After buying a fluid-hauling truck to service oil companies, Hamm made enough money to drill his first well in 1971. In the 1980s, he struck the mother lode, finding a meteor-impact crater near Enid that ended up producing 17 million barrels of oil. Hamm's luck continued as he adopted the new fracking technology while acquiring more than 600,000 acres of the Bakken Shale field in North Dakota. By the second decade of the new century, Hamm's Continental Resources company was producing 700,000 barrels of oil a day—10 percent of the nation's output—with 135 million barrels of booked reserves in the US And the CEO, with three homes and five kids, was now worth billions. The *National Journal* compared him to John D. Rockefeller. In 2012, Republican presidential candidate Mitt Romney named Hamm his energy adviser, while the oilman's two industry fundraisers brought some $10 million into Romney's campaign.

Hamm's good fortune was buoyed by the popular perception at the time that fracking of natural gas was a vital stepping stone toward a clean energy future. In his 2012 State of the Union speech, President Obama spoke of how abundant supplies of fracked gas would last America for a century and result in 600,000 new jobs by 2020. Another Oklahoma-based fracking mogul, Aubrey McClendon of Chesapeake Energy, seized upon the statistic that natural gas emitted half the carbon dioxide of coal-fired plants, and quietly poured $26 million into the Sierra Club's "Beyond Coal" campaign that effectively blocked dozens of new mines. Before long, fracked shale gas from Hamm's, McClendon's, and other companies were providing one-quarter of America's natural gas supply, while the country bypassed Russia as the world's biggest such producer.

But the 2011 Oscar-nominated documentary *Gasland*—with its astounding footage of toxic methane bursting into flame as water gushed out of a family's kitchen faucet—began raising doubts about the booming industry even before the spate of earthquakes in Oklahoma. Then came California's Porter Ranch disaster. Late in February 2016, findings reported in the journal *Science* revealed that a natural gas plume leaking into the Porter Ranch area for four months had not only been invisibly poisoning dozens of families prior to their forced evacuation, but had released some 100,000 tons of methane. This largest methane leak in US history had effectively doubled the emissions rate of the whole Los Angeles basin.

Previously, an industry-funded MIT study (co-authored by Ernest Moniz, Obama's future Energy Secretary) had asserted that "environmental impacts of shale development are challenging but manageable." The federal EPA had estimated low methane leakage rates, and minimized the harm to our atmosphere compared to carbon emissions over a hundred-year period. But this was contradicted in 2011 by two professors at Cornell University, who revealed that fracking natural gas from the Marcellus Shale could cause more global warming than mining of coal. Particularly in the short term—with a pound for pound warming impact 105 times greater than CO_2—the estimated 8 percent of methane that would leak into the air from a fracking well was twice that of a conventional gas well, and a far bigger problem for the climate than anyone thought.

Then, in February 2016, a paper authored by several Harvard researchers appeared in *Geophysical Research Letters*. Based upon satellite data and ground observations, methane leaks from natural gas wells were detected at alarming levels—up by more than 30 percent in the US between 2002 and 2014, causing a huge spike in our planetary atmosphere. Even the Porter Ranch emissions didn't compare to the amount seeping slowly out of millions of pipes, welds, joints and valves. And continuing to frack for oil and natural gas would make it all but impossible for the US to meet its target of a 26 to 28 percent reduction in greenhouse gases from 2005 levels by the year 2025.

One more grim report came out in April, examining the havoc wreaked on various fronts by the more than 137,000 fracking wells permitted since 2005. Besides the 23 billion pounds of toxic chemicals used over a ten-year period, in 2014 fracking released 5.3 billion pounds of methane during simply the completion of wells.

This was followed by the 2016 EPA annual inventory that sharply revised upward its estimates of methane emissions from fracking. Previously, agency data had indicated that the livestock industry was the country's leading source of methane—but no longer. "The oil and gas sector is the largest emitting-sector for methane and accounts for a third of total US methane emissions," the EPA report concluded. Are we surprised that the American Petroleum Institute responded that the EPA's modification "is seriously flawed"?

The sign as you drive across Payne County and into the outskirts of the small town reads: "Cushing Oklahoma—Pipeline Crossroads of the World." After the 1940s, when Cushing's long-booming oil production began to dry up and leave behind miles of empty storage tanks, it became the nation's primary oil depot. Underneath Cushing, thousands of miles of pipelines stretch out toward distribution hubs across the nation. Over the past decade, as US production of fracked oil nearly doubled, Cushing became a crucial distribution center, moving the fuel from Harold Hamm's North Dakota fields to the Gulf Coast refineries. Today, Cushing's tank farms are also the backbone behind $179 billion in West Texas intermediate crude futures and options traded on the New York Mercantile Exchange.

Turning south off the main highway onto Route 18, locally designated as Little Avenue, cylindrical white and brown tanks start lining the horizon for miles. The long row of oil tanks stretch as far as the eye can see, behind barbed wire fences with signs that identify the owners of the fuel supplies: Enbridge Central Terminal, Magellan Midstream Partners LLP, Blueknight Energy Partners, on and on. Where the gravel begins at Texaco Road, a sign says: "Caution: benzene may be present, H_2S poisonous gas." But there doesn't appear to be much security amid the millions of barrels of oil stored in giant tanks, some big enough to house a Boeing 747. Only the federal Strategic Petroleum Reserve contains more.

Coming to Cushing in March 2012, on his first presidential visit to Oklahoma, President Obama addressed workers in a pipe yard about his domestic energy policy. "Over the last three years," he told the crowd, "I've directed my administration to open up millions of acres for gas and oil exploration across 23 different states. We're opening up more than 75 percent of our potential oil resources offshore. We've quadrupled the number of operating rigs to a record high. We've added enough new oil and gas pipeline to encircle the earth and then some.

"So we are drilling all over the place—right now," Obama continued. "That's not the challenge. That's not the problem. In fact, the problem in a place like Cushing is that we're actually producing so much oil and gas in places like North Dakota and Colorado that we don't have enough pipeline capacity to transport all of it to where it needs to go—both to refineries, and then, eventually, all across the country and around the world. There's a bottleneck right here because we can't get enough of the oil to our refineries fast enough. And if we could, then we would be able to increase our oil supplies at a time when they're needed as much as possible.

"Now, right now, a company called TransCanada has applied to build a new pipeline to speed more oil from Cushing to state-of-the-art refineries down on the Gulf Coast. And today, I'm directing my administration to cut through the red tape, break through the bureaucratic hurdles, and make this project a priority, to go ahead and get it done."

The audience cheered. Early in 2014, commercial shipments of crude started moving underground out of Cushing toward Port Arthur, Texas, through the southern leg of the Keystone pipeline. Although Obama would eventually halt the northern extension of the controversial pipeline from Canada following widespread public outcry, the oil from Cushing is still flowing. Since no international boundary is involved, State Department approval wasn't required for TransCanada to move forward.

After September 11th, 2001, Cushing had been considered a prime target for a terrorist attack. Guards were positioned along the vast oil depot's perimeter and new surveillance cameras were put in place by the FBI and law enforcement. All references to tanks and pipelines were removed from the local Chamber of Commerce web site. Simulations of various emergencies were enacted—fires, explosions, hostage-taking. All the while, thanks to Hamm and other frackers, the shale boom brought Cushing's capacity to a record 60-million-plus barrels by the spring of 2015.

And then, from underground, a different threat emerged. At a local diner, a fellow in coveralls chats about it with a visitor at the lunch counter. "Yeah," he says in response to a question, "I think there's some fracking going on around here." He himself lives in nearby Medford, where one night "you could hear a rumbling like a train in the distance. In all my years growing up, I never knew of any earthquakes in Oklahoma until now."

In the fall of 2015, more than a dozen quakes registering at 3.0 or higher—with the biggest at 4.5—struck within a few miles of the pipeline crossroads. A dilapidated building in downtown Cushing collapsed. The Oklahoma Corporation Commission ordered volume reductions and even shutdowns at five wastewater disposal wells in the vicinity. The EPA urged the commission to "implement additional regulatory actions."

The earthquake that rumbled like a train struck Medford in mid-November 2015, at 4.7 the strongest in Oklahoma since 2011, and was felt in seven other states. A year later, Cushing recorded an even larger quake at magnitude 5.0, causing the regulators to shut more disposal wells and reduce the volume of others. Should a rupture happen, each tank in Cushing is surrounded by a clay-lined berm aimed at controlling an oil spill. But as an operator for the Unbridled Energy company acknowledged, his 18 tanks were not built to withstand a serious quake. According to petroleum geologist Bob Jackman, should a big one occur, "Cushing would be the world's biggest fireball." And that would send a mighty tremor across the American energy market.

"There are people in Cushing who are terrified," Angela Spotts was saying. "Their homes are crumbling. But they're uncomfortable about suing. I really think our state is a sacrifice zone. Harold Hamms, David Borens, Boone Pickens—they do not care."

Originally from Tulsa, Spotts and her husband had moved eight years ago to a rural area just outside Stillwater, home of Oklahoma State University. The earthquakes started in 2012 while her terminally ill mother was living with the family, and "we couldn't figure it out for a long time." But as time went by, Spotts began to suspect the trouble was related to Devon Energy.

That company was run by Larry Nichols, who, as a Princeton graduate and ex-Supreme Court clerk, was Hamm's opposite in background. Nichols had "urged his family's Oklahoma energy company to buy Mitchell Energy after he noticed that its natural gas output was climbing because of fracking," according to Jane Mayer's book *Dark Money*.

"That's when fracking came to our door," Spotts recalled. "They'd show a map basically blood-red, saying 'We're gonna put a well on every square mile we can.'" Big trucks soon filled the roads. "And OSU was able to keep $100,000 in earthquake damage on a building out of the news for a year," Spotts said. Meanwhile, she noted, numerous events on the OSU campus featured Devon Energy spokespeople.

The Spottses and their neighbors soon began to suffer mysterious ailments. As neighbors complained that their water wells were going bad, Spotts' husband had 15 nosebleeds in a week, while their family dog got sick and had to be put down. "Living near all this [fracking] has had much more of an impact health-wise than people give credit to," she said. Angela formed an organization called Stop Fracking Payne County. The back of the group's T-shirts read: "Don't underestimate the persistence of our red dirt resistance."

But Spotts found it an uphill fight. Although 700 people showed up at one town hall meeting and activists gathered regularly in her living room, Spotts learned that some were threatened with losing their university jobs while retired teachers found new employment driving as escorts for the rigs. Other citizens worried that, with oil prices having already plummeted from $100-a-barrel to around $30, negative comments about the industry might cause it to pull out of the state.

Oil, and more recently natural gas, run deeply in the bloodline of Oklahoma, all the way back to statehood in 1909. An iron oil derrick stands tall on the south side of the state capitol grounds in Oklahoma City, while in Tulsa the 75-foot-tall Golden Driller is the fifth highest statue in the US. The Oklahoma Corporation Commission, charged with regulating the industry, was chartered long ago to protect and preserve it. About one quarter of the state budget emanates from direct taxes on the drillers, who also pay royalties to farmers and homeowners. An estimated one-third of Oklahoma's economy and one in five jobs depend upon fossil fuels.

It's no accident that the US Senate's leading climate change denier, Oklahoma Republican and Environment Committee chair James Inhofe, receives more oil and gas contributions than anyone else in Congress (nearly two million dollars over the past two decades, with Devon Energy topping his donor list). Inhofe is the source of some of Washington's most unhinged rhetoric and grandstanding on the climate crisis, declaring, "Global warming is the greatest hoax ever perpetrated on the American people," and arguing that the atmosphere naturally fluctuates between warming and cooling periods. As Washington experienced one of its snowiest winters ever in 2010, Inhofe enlisted his daughter Molly, her husband, and the senator's four grandchildren to erect an igloo at a capital intersection, dedicating it as "AL GORE'S NEW HOME" on one side and "HONK IF YOU LOVE GLOBAL WARMING" on the other. Later, as part of a rambling speech to fellow senators during which Inhofe disputed scientific evidence about 2014 being the warmest year in recorded history due to climate change, the Oklahoman held up an icy globe and said: "I ask the chair: do you know what this is? It's a snowball!" He then threw his "proof" into the hands of a Senate page.

Six months later, shortly before Pope Francis issued his encyclical about climate change, Inhofe addressed the right-wing Heritage Foundation. He pointed to the sky, asserted "God is still up there," and added that the pope should "stay with his job, and we'll stay with ours." As petroleum geologist Bob Jackman puts it, Senator Inhofe is "a big-time Sunday Christian and a Monday thief."

Meanwhile, Oklahoma's junior senator, James Lankford, sticks to the economic/political argument against climate action. When President Obama rejected the northern leg of the Keystone XL pipeline late in 2015, Lankford said: "This is what happens when the federal government gets to choose projects based on the preferences of a few elites in Washington, DC, rather than the free market and the preferences of the people. . . . We have truly lost basic constitutional federalism."

Mary Fallin, Oklahoma's governor since 2010, received more than $800,000 from oil and gas interests during two terms as a US Congresswoman, a trend that continued with her 2014 gubernatorial re-election campaign, to which oil and gas producers were the biggest contributors. In return, she signed a bill that year to protect the industry from local government ordinances, then proceeded to slash funding to numerous state agencies, including the one in charge of monitoring drilling activities and earthquakes.

"It's important that we study this issue and have sound science that can inform decisions," she'd previously said about the surge in quakes. But it wasn't until September 2014 that Fallin finally formed a coordinating council to study seismic activity. In August 2015, she conceded a direct correlation between the fracking surge and the earthquakes, but suggested that people suffering damage call their insurance agent "and see what types of products are available." The year 2015 saw 907 quakes of 3.0 or higher recorded, almost 50 percent more than 2014, whose 535 such quakes were more than the number in the past 35 years combined.

Scott Pruitt, elected Oklahoma's attorney general in 2010, started out by folding his office's Environmental Protection Unit and replacing it with a Federalism Unit that proceeded to initiate Pruitt's legal challenges against the EPA. He was among the primary creators of what's called the "Rule of Law" campaign, bringing together his Republican counterparts from across the country in defiance of the Obama EPA's regulatory agenda. This eventually resulted in a lawsuit filed by 28 states against the administration's rules. "I don't think there is anything secretive in what we've done," Pruitt has said. "We've been very open about the efforts of my office in responding to federal overreach." A Pruitt letter to the EPA, accusing the agency of overestimating air pollution from fracking, was revealed by the *New York Times* to have been written by attorneys representing Devon Energy.

Harold Hamm co-chaired Pruitt's successful 2013 re-election campaign. The following year, Pruitt got together with the Domestic Energy Producers Alliance—chaired by Hamm—in suing the federal government, concerning plans to determine the status of 251 animal species for the Endangered Species List. Hamm was particularly concerned that the habitat for the prairie chicken overlapped "some of the most promising land for oil and gas leases in the country."

The 48-year-old Pruitt was chairing the Republican Attorneys General Association when Trump chose him to head the EPA in December 2016. Hamm was thrilled. Trump, in Hamm's words to CNBC, "continues to pick awfully good candidates for all the Cabinet posts. He's following through with what he told the American people."

Ken Cook, head of the Environmental Working Group, called it "a safe assumption that Pruitt could be the most hostile EPA administrator toward clean air and safe drinking water in history." Not to mention rules on climate change, of which Pruitt flat-out lied early in 2016 in an article for the *National Review*: "Scientists continue to disagree about the degree and extent of global warming and its connection to the actions of mankind. That debate should be encouraged—in classrooms, public forums, and the halls of Congress."

In January, at Pruitt's Senate confirmation hearing for the EPA post, he refused to say whether he would recuse himself from taking part in fourteen yet-pending lawsuits against the agency's Clean Power Plan.

Even more alarming, Pruitt wouldn't commit to retaining a decades-old federal waiver allowing progressive places like California to establish stronger emission standards on motor vehicles than other states have. Since California did so in 2009, its emissions have fallen by nearly a third and more than a dozen other states have adopted similar standards. It seems that Pruitt is all for "state's rights"—except when they get in the way of fossil fuel production.

The Senate refused to wait on confirming Pruitt until a court order forced his office in Oklahoma City to turn over over 7,000 pages of internal emails to the Center for Media and Democracy. When the contents were revealed several days later, they confirmed that Pruitt worked hand-in-hand with the fossil fuel industry that his state was supposedly to regulate. His office was particularly chummy with Devon Energy, which not only held regular meetings with Pruitt but often drafted letters for his signature to the federal government opposing new regulations.

Once anointed, Pruitt wasted no time in flexing his muscles. He told the *Wall Street Journal* that withdrawing both the Clean Power Plan and the Waters of the United States rule promulgated by Obama would be only the start of the process. Speaking to a national gathering of conservatives, Pruitt said those who think the EPA should be eliminated entirely were "justified" given what the previous administration had done.

Meanwhile, the energy industry's domination of Oklahoma continues unabated. Philip Anschutz, a large portion of whose fortune derives from oil and gas, is the owner of *The Oklahoman*, the state's leading newspaper. Fossil fuel companies donate millions to the arts, community economic development, and sports franchises like the NBA's Oklahoma Thunder (partly owned by Chesapeake Energy's Aubrey McClendon before his suicide). The companies also fund most of the election battles for seats on the three-member Oklahoma Corporation Commission. In 2011, one commissioner resigned to go to work for Hamm's Continental Resources. Another, Dana Murphy, has pointed out that the state constitution was written to protect the underworld over the surface world. She apparently saw no irony in that statement.

It's been estimated that the University of Oklahoma (OU) and Oklahoma State University (OSU) are now 85 percent funded by wealthy private donors, especially those connected to big fossil fuel companies. "The university presidents have really become telemarketers," says watchdog Bob Jackman.

In a letter written by OU President Boren in March 2013, the incestuous relationship couldn't have been more obvious. It was addressed to Mike Smith, executive director of the Interstate Oil and Gas Compact Commission, concerning "your participation in the briefing for staff members of the Bloomberg Foundation, the Environmental Defense Fund, and the Mitchell Foundation on new technologies in the oil and gas industry, including horizontal drilling and fracking." Boren, himself a trustee of the Bloomberg Foundation, described this as "an opportunity to demonstrate that regulation is best handled at the state level as opposed to the federal level. He also suggested that Continental Resources could serve as a good private sector model for responsible exploration and operations."

At the ensuing "technical briefing" on the OU campus, Boren offered the welcoming remarks and then introduced Harold Hamm, whose topic was "How Continental Drills Safely." Following the second panel on "Evolution of Oil and Gas Law—Origination and Transfer," participants were invited to lunch at Boyd House, the president's on-campus home.

While Continental Resources, Devon Energy, and Chesapeake Energy dominate the same block of downtown Oklahoma City, their influence fans out to more than one school. Indeed, three of the four major universities in Oklahoma have presidents who have doubled as paid directors of a fossil fuel company.

Burns Hargis, formerly on the Chesapeake board, has been the president of OSU since 2008. That same year, Devon Energy donated

$1 million to OSU toward establishing a faculty chair dedicated to the school of geology's basin research. The university announcement noted that the company had already invested heavily in the school, with gifts totaling over $2.3 million for the previous four year period, including financing of the Devon Energy Geology Laboratory. More recently, in January 2016, OSU President Hargis hosted the National Engineering Forum, with Devon Energy's new CEO David Hager delivering the keynote speech.

Another major contributor (and OSU graduate), T. Boone Pickens, has the football stadium named after him. Pickens once bragged he was heading to Washington to sell natural gas to Obama as a "bridge fuel." In 2014, he opined of the earthquakes: "If there weren't a bunch of these silly seismometers, people wouldn't know we were shaking." The OSU Alumni Center is named ConocoPhillips. Governor Fallin graduated from the university. She's publicly praised the Charles Koch Foundation for giving $105,400 to the Oklahoma State University Foundation between 2005 and 2014.

Robert Henry, president and CEO of Oklahoma City University since 2010, received compensation as a Devon Energy board member of $336,015 in 2013. Steve Agee, dean of the business school, is the former president of the Oklahoma Energy Resources Board, the state's version of the American Petroleum Institute. John Richels, Devon CEO before retiring in 2015 (and also a paid director of the Trans Canada pipeline company) remains on the board of trustees of Oklahoma City University. Masters programs in energy there are directly tied to Devon Energy and Chesapeake Energy, with most graduates finding employment with one of the two companies. The school's Agee Economic Research Institute does paid research for Devon, which the company has used to lobby the federal government. Some professors become salaried consultants during their summer breaks.

Despite this oil industry largesse, since 2008 Oklahoma has led the nation in budget cuts to education. That is primarily because the massive tax breaks given the oil-and-gas industry for "unconventional drilling practices," specifically horizontal related massive fracking, has depleted the state's tax coffers. Energy companies' annual tax rate stood as low as 1 percent during the shale boom. For fiscal 2015, they went up to 2 percent. By contrast, North Dakota taxes the industry at seven percent, and has been able to increase education funding more than any other state during the same period. Had Oklahoma followed suit, this would have recouped between $300 million and $400 million a year of lost revenues. As it stands, a thousand school jobs are at risk and 100-plus school districts are forced to contemplate shorter weeks or school years. Oklahoma also ranks near the bottom nationally in public health.

Mark Davies, who's taught at Oklahoma City University (OCU) for 20 years, worked his way up to become dean of arts and sciences by 2009. But as he watched his school become more and more tied to the fossil fuel industry, Davies became a climate activist. He was among the chief organizers of the People's Climate March in November 2015, moving down the same street where Continental, Devon, and Chesapeake have their headquarters. Warned earlier that if he were ever to protest at Devon Tower, it would be the end of his time as dean, Davies decided to step down and take two other positions, despite a $30,000 pay cut. He remains at OCU as founding director of the World House Institute for Social & Ecological Responsibility.

Davies is the exception. Most university administrators in Oklahoma find a way to accommodate themselves to the conflicts of interest, the research for hire, and the stained corporate patronage.

It is unlikely you've ever heard of the Interstate Oil and Gas Compact Commission (IOGCC). Established by Congress in 1935, when the domestic oil industry was in difficult straits during the Great Depression, it represents 38 states as a quasi-governmental "shadow lobby," whose main aim is to restrict oversight by the federal government. The IOGCC, one third of whose 495 individual members have direct ties to the fossil fuel industry, uses a loophole in the Lobbying Disclosure Act to describe such efforts in Washington as "education." Nor does the IOGCC need to comply with open records laws.

The IOGCC's office, once inside the state capitol building in Oklahoma City, today is located on property next to the governor's mansion. That's convenient for current Governor Mary Fallin, the organization's chair in 2011 and again in 2016. Michael Teague, Oklahoma's first-ever secretary of energy and environment, is one of two vice chairmen. Harold Hamm has long been one of the leading IOGCC members.

In 1978 the Justice Department's antitrust division under President Carter told a congressional subcommittee that the IOGCC had outlived its original function. "If so much activity should continue to be taken up with what is essentially lobbying work, it would seem inappropriate for it to have the special cachet of congressional approval," said Donald Flexner, who headed the antitrust energy office. Such approval was withdrawn, but that didn't stop the IOGCC from continuing to function over the past nearly 40 years, its annual budget funded partly through an excise tax on petroleum—which means that, when oil and gas production goes up, so do the group's finances.

In 1999, the nascent stage of hydraulic fracturing, the IOGCC passed a resolution in support of a bill introduced by Senator Inhofe to exempt fracking from the Safe Drinking Water Act. This didn't happen until six years later, when a similar provision known as the "Halliburton loophole" (then-Vice President Cheney's former oilfield services company) became part of the 2005 Energy Policy Act as the Bush administration began its second term. The frackers also received an exemption from the Resource Conservation Recovery Act, meaning the wastewater now responsible for causing earthquakes couldn't be regulated as hazardous waste. In its newsletter, the IOGCC announced proudly: "Congress passes IOGCC's legislative fix for hydraulic fracturing," bringing "several years of hard work by the Commission to fruition."

This scuttling of the EPA's ability to regulate under existing laws meant that fracking could take off—and it soon did, to the delight of Hamm and his cronies. So did fracking's less desirable results: contamination of drinking water, air pollution, and increasing emissions of climate-changing methane. The disastrous methane leak from Southern

California's Aliso Canyon natural gas storage facility at Porter Ranch in 2015–16 might have been prevented had the IOGCC not been allowed to undermine federal law. A working group to study underground storage concerns was subsequently set up by the IOGCC, but its opposition to EPA oversight didn't change.

In 2011, the IOGCC established FracFocus in tandem with another Oklahoma City-based group, the Ground Water Protection Council, hailing its database on chemicals injected into the ground during fracking as an example of transparency that precluded any need for federal involvement. But critics contend that corporate withholding of information for proprietary reasons make the database flawed and inaccurate.

Also in 2011, IOGCC deputy director Gerry Baker—once a PR man for the Kerr-McGee Corporation, another Oklahoma-based energy giant—gave a presentation at the annual meeting. "When people do not get all of the information, or information is too technical, they begin to fill in the holes with what they can imagine. The perceived risk, even if it isn't a reality, makes hydraulic fracturing an emotional issue."

In 2013, the IOGCC collaborated with then-House Speaker John Boehner's office in pushing for legislation to prevent the Bureau of Land Management (BLM) from regulating fracking on public lands. Two years later, a US district judge cited the Halliburton loophole in blocking the BLM's still-pending regulations.

In 2015, the IOGCC's annual conference convened in Oklahoma City, sponsored by some of the world's biggest fracking outfits. "We see talking points being passed directly to regulators, who then go testify in Congress," according to Greenpeace-USA researcher Jesse Coleman. "They also get their governors to send letters to regulatory authorities that copy the exact language from the IOGCC." The organization, Coleman concludes, is "a mechanism for really subtle corruption."

The Western Fuel Alliance, a fracking industry-funded front group, gave a presentation at the IOGCC's annual meeting in Denver in May 2016. It described a new "Keep it in the Ground" citizen's movement that had been showing up at BLM lease sales across the West. In Colorado, the alliance spokesman said, about 250 protesters had "tried to storm the Holiday Inn at one point to block access to the auction." So the alliance was "planning some counter-efforts," pushing the BLM and Congress to "get rid of the circus" and move to online auctions.

The industry mind-set has little tolerance for opposition. In Oklahoma, questionable things have happened in the past to individuals who speak out against the big energy companies. Remember Karen Silkwood, who, after raising safety concerns about the Kerr-McGee plutonium plant where she worked, was found dead in what police alleged to be a single-car accident in 1974. Recently, when activists came together to protest expanding the Keystone-XL tar sands pipeline south from Cushing, Moriah Stephenson recounts, "Information emerged that the Oklahoma City police had been working with the Joint Terrorism Task Force, and companies like Devon Energy and Trans-Canada, to use terrorist charges very intentionally as a tactic to repress dissent."

The number of people on Homeland Security's terrorist watch list came to include dozens of citizens protesting along the pipeline route. Stefan Warner, a young activist with the Oklahoma City-based Great Plains Tar Sands Resistance Group, has been arrested three times. Warner was accused by police of making a terrorist threat after saying he would "do anything to stop this pipeline." A judge dismissed that case, but when he accidentally bumped into a firefighter at another protest, Warner got arrested for alleged assault-and-battery on a public official, which carries a potential five-to-seven-year prison sentence. His third arrest, for hanging a banner off the Devon Tower downtown, resulted in a "terrorism hoax" charge that's still pending, though the statute of limitations is set to expire. Meanwhile, all of Warner's Freedom-of-Information-Act requests have been denied.

Some observers found it curious timing that, shortly after the news broke in 2015 about the secret meeting between Hamm, Boren, and OGS seismologist Holland, a video was leaked to local media that showed members of a university fraternity singing a racist chant on a bus. When

OU President Boren expelled the students and came out looking like a hero, this quickly overshadowed reports of the oil-and-gas industry's questionable ties to the university.

When it comes to the earthquake measurements, community organizer Angela Spotts says, "There used to be a joke: Hamm wouldn't allow fours." That was the magnitude level at which scientists concurred that action needed to be taken to curtail the wastewater injection triggering the quakes. But suddenly, according to Spotts, "the Geological Survey was taking many measurements that registered in the low magnitude-fours and turning them into M-3.9s."

That's not surprising, given that the OGS was originally written into the state constitution as a geological research entity meant to service the oil-and-gas industry. Graduate assistants at Oklahoma University sometimes receive free tuition in exchange for testing rocks from deep underground for the presence of petroleum.

Despite this, observes OGS hydrogeologist Dr. Kyle Murray, "Not many would disagree at this point" about the links between fracking and the quakes. When fracking increased in the northern Mississippian and Arbuckle Group fault zones, Murray said, so did the volume of seismicity. Murray says "probability would suggest there is the potential for five or six-point earthquakes. We have the faults and some are oriented in a direction that's prone to fail." (An article in *Science Magazine* noted that an area near Oklahoma City is even capable of seeing a 7.0).

But rather than pay for transport of its wastewater byproducts down the road to technically safer disposal wells, the industry circulated word that small quakes actually prevent a larger one. The frackers also said the problem could be mitigated if stronger material was used in the building of houses, such as steel. "There's a hysteria that needs to be brought back to reality," asserted Glen Brown, a geologist with Hamm's Continental Resources company, in 2014, adding that "these are light [earthquakes] and will not cause any harm."

But retired oil and gas operator and petroleum geologist Bob Jackman is not buying the industry spin: "They're doing the classic propaganda—reframe the issue, make it more complex to where people can't understand it. The message has not been clear to citizens that it's only about ten of our 77 counties that are subject to and getting the earthquakes. We have 44 formations in Oklahoma that produce oil and gas. A particular one that produces enormous quantities of brine and water, called the Mississippi line, runs across the north-central part of the state. We can live with the amount of brine byproduct if it's eight or ten barrels for every barrel of oil. But when they started doing horizontals [fracking], now we've got 40,000 to 60,000 barrels of brine a day per well. When you put that much through disposal wells onto these somewhat fragile geological structures, it will cause either slippage or the sheer weight brings displacement, which also creates earthquakes. They just have to shut down about 200 of these monster disposal wells [of some 3,300 in Oklahoma] in those ten counties, which may mean cutting out five to seven percent of daily production. A small price to pay, but this state has clearly chosen profit over people's safety. This is a repeat of the asbestos and tobacco stories."

As *Scientific American* reported in its July 2016 issue, "Geologists have known since the 1960s that pushing fluids into the ground can set off quakes." After residents near the Rocky Mountain Arsenal near Denver experienced over 700 small to moderate-sized quakes following the drilling of a deep well at the chemical weapons facility, local geologist David Evans's paper concluded that "a stable situation is being made unstable by the application of fluid pressure." In 1966, the Army shut down the disposal well—only to find that the earthquakes continued "and even grew stronger as pressure from the injections propagated belowground, propagating new faults and disturbing them."

Despite the fracking boom, it hadn't been an easy time for Harold Hamm. As oil prices plunged, shares in his Continental Resources company fell more than 60 percent between late 2014 and March 2016. His net worth, as ranked by *Forbes*, tumbled from $18.7 billion in 2014 to around $5.1 billion, with banker-oilman George Kaiser supplanting

Hamm as Oklahoma's wealthiest individual. Hamm's recent divorce settlement further depleted his worth, peeling off nearly $1 billion for Sue Ann, his wife of 25 years, one of the biggest divorce sums ever awarded.

But Hamm's turn of fortune didn't stop then-presidential contender Donald Trump from recruiting the oil mogul to shape his energy policy. The two men first became acquainted late in 2012, when, at a private meeting, Trump took notes on Hamm's fracking expertise and (according to Hamm's recollection) tossed out the notion that he might run for high office. Trump proceeded to give Hamm some of his branded ties, one of which the oil billionaire sported for a magazine cover, reportedly to Trump's delight.

After Hamm endorsed Trump's candidacy in April 2016, calling him the "business leader's candidate," Trump in turn dubbed Hamm "the king of energy." Rumored to be the nominee's choice for energy secretary should he win, Hamm entered the national spotlight when asked to speak at the Republican Convention. "Climate change isn't our biggest problem," Hamm told the Trump faithful gathered in Cleveland. "It's Islamic terrorism." He went on in the same orange-alert vein: "We can double US oil production again and put America in a global league of its own. Every time we can't drill a well in America, terrorism is being funded. Every onerous regulation puts American lives at risk." Masquerading as an energy security speech, it was a greedy pitch as shameless as the GOP candidate himself.

After Trump's victory, Hamm would turn down the president-elect's offer to be the next Secretary of Energy. This may have something to do with the post-election surge in Continental Resources stock, which garnered Hamm $3 billion in under three hours following OPEC's announcement to finally cap production at 32.6 million barrels a day. Hamm's net worth soared again to around $14.7 billion—making him *Forbes'* 60th wealthiest person on the planet, with a fortune almost four times that of Trump.

In the meantime, Oklahoma's response to the earthquake swarms has been the least proactive of any state faced with such a dilemma. Arkansas imposed a fracking wastewater disposal moratorium across much of the state in 2011, and the quakes stopped. Kansas shut down wells in the southern part of the state, which is geologically the same as northern Oklahoma. After an unusually strong tremor near the Dallas-Fort Worth Airport in 2012, then-Texas Governor Rick Perry approved $4.5 million for a comprehensive study, compared to the initial $50,000 allotted to Oklahoma's Coordinating Council on Seismic Activity. A University of Texas-Austin study published in the spring of 2016 in *Seismological Research Letters* revealed that nearly nine-tenths of the 162 earthquakes (M-3 or higher) experienced in Texas over the past four decades may have been caused by oil and gas activities.

Recently, as Oklahoma went from two quakes a year in 2008 to *two a day* by 2016, the pressure finally forced some movement by state officials. In January 2016, a series of earthquakes caused power failures, toppled columns and stripped off brick facades in Edmond, an upscale suburb of Oklahoma City that's home to a number of the state's political and financial bigwigs. This was followed a week later by a pair of quakes, registering 4.7 and 4.8, in rural northern Oklahoma directly under a major oil-and-gas production area.

By March, already almost 160 quakes had been registered. As her state kept rocking, Governor Fallin finally allotted $1.4 million to the OGS and Oklahoma Corporation Commission [OCC] to shore up their monitoring networks, and the OCC demanded that well operators in central Oklahoma reduce the volume of injected wastes to 40 percent below 2014 levels. Previously, regulation had been confined to a handful of wells at a time. The Oklahoma Supreme Court was also forced to act, agreeing to review a trial judge's ruling against a class action suit demanding corporate accountability for earthquake damage. The CBS news program *60 Minutes* brought national attention with a segment on the earthquakes in April 2016 but, according to Bob Jackman, "like many reports, it failed to disclose Oklahoma's 100 to 150,000 homes with quake damages." That became harder to ignore in early September, when a state record-tying 5.6-magnitude tremor forced Governor Fallin to declare a state of emergency for Pawnee County—and pushed the Oklahoma Corporation Commission to shut down three dozen wastewater wells across some 500 square miles. While Pawnee residents filed a class-action lawsuit against 27 oil and natural gas companies, an analysis of USGS data revealed that the rate of earthquakes linked to fracking had dropped significantly since the new regulations cutting back on wastewater injection went into effect in May.

Whether this will significantly shift the ground under Oklahoma's corporate and political elites remains to be seen. For now, Oklahoma retains the dubious distinction of being earthquake capital of the lower 48 states. And the kings of fracking remain defiant. Harold Hamm still talked trash about renewable energy, telling the *National Journal* that clean energy sources don't deserve federal subsidies—partly for aesthetic reasons. "I frankly don't like to see a wind turbine," Hamm mused. "Once they're there, they haunt you. That's your viewshed. That's what you look at. All those things standing out in the distance, we have them all over Oklahoma. And it doesn't look very good."

For some reason, the fracking towers, waste disposal wells, and pipelines that mar the Oklahoma "viewshed" don't haunt Hamm the same way. In any case, Hamm is determined to keep tilting at wind turbines, even though Oklahoma is currently the fifth-largest wind-energy producer in the US. "We do have a lot of wind farms and they're growing quite rapidly, typically outsiders coming in," says Bob Jackman. "But the mind-set [of the fossil fuel industry] is they want to take away any tax credit from wind farms, and hit with a surcharge anybody who puts a solar panel or windmill on their property."

While Hamm argues vociferously against the wind energy tax credits, asserting they are "special treatment," he insists that government subsidies for his company are perfectly fair. "I need them to refine our oil," he states. Hamm also dismisses the role of renewable energy sources in battling climate change. "Overpopulation—that probably hurts the environment more than anything," he also told the *National Journal*.

In December 2015, the *DeSmogblog* web site published a document authored by the Public Relations Society of America, revealing the "Miracle of American Oil" campaign to push Congress and the White House to lift the ban on crude oil exports. The effort was spearheaded by Continental Resources' own PR team, which kicked things off with a five-star dinner in Washington attended by a number of influential financial journalists, as well as Hamm himself. According to Continental, the campaign "served as a catalyst to correct public misconceptions." At the same time, Hamm's company set about to "change the vernacular from fracking to horizontal drilling," a language shift which Hamm emphasized five times during congressional testimony in January 2014 and four more times in an opinion piece for *Forbes*.

Hamm gave a PowerPoint presentation in June 2015 to the annual Energy Information Administration conference in Washington. It was titled "Rigs, Rednecks, and Royalties." In his talk, Hamm described the oil-producing regions of the country as "Cowboyistan." This is clearly the real America, in the minds of men like Harold Hamm. Wind and solar energy might be fine for places like California, but the black gold beneath the ground is what made states like Oklahoma and empires like Hamm's. Now his man Scott Pruitt is going to be running the federal EPA.

Chapter 7

Peabody Energy's Global War on Children

Go to the web site of the world's largest publicly traded coal company, and the faces of children abound. Peabody Energy is fond of highlighting how employees in its St. Louis headquarters participate in a Junior Achievement day at local elementary schools, bringing a curriculum that encourages students "to own their economic success by fostering work-readiness, entrepreneurship and financial literacy." Meanwhile, the Peabody Energy Leaders in Education program honors the city's teachers with financial awards up to $5,000. And another group of employees walks a mile to help the Make-A-Wish Foundation raise funds for children with life-threatening medical conditions.

Peabody, which does business on six continents and in over 25 countries, claims that its concerns for children go beyond the St. Louis school district. In November 2013, the company bought a double-page advertisement in a special edition of the International Energy Agency Journal that displayed a color photo of a turbaned man riding a bicycle down a dirt road in India, with his son on the back of the seat. "Our Children Turn to us for a Brighter Energy Future," the headline read.

"Too many of the world's children have no modern energy in their lives—no light to read by, no computers for school work, no digital devices to bring their communities into the 21st century," the ad's text began. "Advanced coal is changing all of that. Coal has been the world's fastest growing major fuel of the 21st Century. And coal is expected to pass oil as the world's largest energy source in coming years." The ad urged the world's energy ministers to "emphasize that today's advanced coal generation is fast-growing, abundant, inexpensive and clean." Peabody's campaign "to eliminate global energy poverty" through burning thousands of BTU's more of its fuel had begun.

Nothing was said about the fact that one-third of all the planet's carbon emissions come from burning coal - some *eleven billion* tons of CO_2 a year. To hold global warming to even two degrees Centigrade, 80 percent of the world's coal reserves need to stay in the ground. And, of course, nothing was said of the respiratory ailments, cancers and other diseases stemming from carbon pollution, or the rapid rise of climate catastrophes—all of which take a particularly severe toll on the young.

Barney Bush, a nationally renowned Shawnee/Cayuga poet and indigenous activist, sits in the study of his cabin in the woods of southern Illinois, reading aloud from a yet-untitled work that he's completed over the course of a year:

A strip mine holocaust operates below
the hill gouging deeply into the
earth eradicating old village sites
and graveyards . . .
Early December and we still have
not had the hard freeze that shuts down
all but the silence . . .
Everyone
every thing knows something is
wrong

too close to catastrophe
out of balance
when hearts turn to dust
and horizons glow portentously
long after the sun has set . . .

Bush, just turned 70, grew up here. Later, over a country breakfast he's made of organic ham, eggs, and homemade biscuits, he says, "Back when I was about 13, the coal mines came in over here and conned our relatives out of the land, by claiming that 'eminent domain' was the law. So I got to see the places where I grew up turned under a strip mine shovel—where we rode our horses, the fields where people planted their corn. The creeks we swam in turned orange and the animals wouldn't drink the water. I still have brochures from Peabody [promising] they'd fix up a strip-mined area and make it into a 'paradise.' I knew then, 'This is bullshit.'

"So some of my buds and I started making night raids on their equipment houses—that bookcase in the corner is made out of some of the dynamite boxes we took—and we'd set fire to a lot of things, too. But it didn't stop anything. I realized that no matter what we did, money and the power of wealth was the number one factor. And I had to learn how to live with that.

"I think I'm a poet because I love my homeland," Bush adds quietly.

He began talking about the changing weather patterns, the disappearance of winter and the daffodils blooming early this year, at the end of January. "But around here, the phrase that comes first before water or air or earth is, 'Yeah, but we need them jobs.'"

Joining us at the breakfast table is Georgia de la Garza, founder of a local environmental group called Shawnee Hills and Hollers. She remarks on how the migratory patterns of the birds have shifted and how the most recent winter was marked by terrible flooding. "A lot of people still say, 'climate change, that's a liberal thing, this is maybe just what the earth does every once in awhile.' Yeah, like wipe out civilization?"

Not long ago, de la Garza posted on her Facebook page a superimposed image of a man's face over a little boy's body, wearing scuba equipment and floating across an impoundment pond. "On Vacation in Rocky Branch—Peabody's Refuse 2 Dam—Saline County, Harrisburg, Illinois," it was captioned, and signed: "Love, Gregory Boyce." "That's where I think he belongs," de la Garza says, "in the middle of what his company has created, and still is." Gregory Boyce is the CEO of Peabody Energy.

Peabody Energy and its top executives have found particularly insidious ways of justifying their actions. But before we get into that, a brief company history is in order. Peabody Coal began as a delivery service in 1880s Chicago, and by 1895 founder Francis Peabody sank the first coal mine of his own in southern Illinois. As World War I approached, Peabody was becoming a major coal supplier, servicing electric utilities and other large customers. The company was bought by Kennecott Copper, then sold to Newport Mining, until Lehman Brothers finally spun it off in a leveraged buyout.

In 1957, Peabody moved its headquarters to St. Louis, a hub for several other coal companies. Peabody became known as Big Coal's battering ram, leading the industry offensive against the Clean Air Act in 1970, spearheading the fight against acid rain regulations in the 1980s, and obstructing later efforts to tighten restrictions on mercury emissions. Ira Engelhardt, then Peabody chairman, advised the new Bush-Cheney administration on energy policy in 2001, and—joining forces with Exxon—convinced the president to abandon his campaign pledge to regulate power plants' CO_2 emissions from power plants and to back away from the Kyoto Protocol international climate-change agreement.

When it came to climate change, Fred Palmer was the Peabody point man. After starting out as chief counsel at a Washington law firm, Palmer became the top lobbyist for the Western Fuels Association, a coalition of coal-burning utilities. By the early 1990s, there was an increasingly strong scientific consensus that global warming was attributable to human activity. But before the 1992 Earth Summit in Rio de Janeiro, Palmer worked closely with the Information Council on the Environment, an industry-funded group, to find ways to "reposition global warming as theory (not fact)." They aimed to "use a spokesman from the scientific community" to bolster their credibility. But not all of their messages sounded scientific. One stated, "Whoever told you the earth was warming—Chicken Little?"

In 1998, soon after the Kyoto accord, Palmer formed the Greening Earth Society. Its message, promulgated via full-page ads and a feature-length documentary, proposed that industrial greenhouse gas emissions would actually bring about flourishing plant life. Neatly boxed videotapes and literature got delivered to every office on Capitol Hill.

"Every time you turn on your car and burn fossil fuel and put CO_2 in the air, you are doing the work of the Lord," Palmer proclaimed publicly. "It is the ecological system we live in." Another time, he said: "We have a sustainable energy delivery system in the United States—it's called coal-fired electricity. . . . We're 100 percent coal. More coal. Everywhere. All the time." And proudly to PBS: "We are at the top of the food chain as far as CO_2 emissions are concerned."

Carrying such coal-friendly credentials, Palmer came to Peabody Energy in 2001 as Senior Vice-President of Government Relations. As Peabody's top lobbyist, he soon was in wider corporate demand, tapped to be a board member of the US Chamber of Commerce, chairman of the National Coal Council's policy committee, chairman of the climate change task force for the National Mining Association, and two-term chair of the World Coal Association. Everywhere you looked, there was Fred Palmer.

At his side was Craig Idso, hired around the same time as Peabody's chief environmental adviser. Idso had previously run the Center for the Study of Carbon Dioxide and Global Change, which held that rising CO_2 "brings growth and prosperity to man and nature alike . . . the Greening of Planet Earth."

Palmer and Idso both reported to Peabody CEO Gregory Boyce. Trained as a mining engineer, Boyce graduated from the Harvard

Business School and began his career with Standard Oil of Ohio, jumping to the mining industry where he was finally hired as Peabody's chief operating officer in 2003. Three years later, he was CEO, soon adding the title of chairman.

Boyce was a round-faced, square-jawed man whose management style appealed to Wall Street, a rising star in the coal industry when energy consumption was approaching new highs. This was despite a 2009 *Newsweek* report that found Peabody's weak compliance with environmental regulations placed it dead last among 500 big companies. By that year, Boyce had devised a new strategic plan that would remove the company further from US environmental controls: Peabody decided that its future business opportunities would be primarily in Asia and the Pacific Rim.

Peabody was already the only US-based coal company in Australia. In 2006, the year Boyce assumed the top corporate position, he brokered a $1.5 billion acquisition of Australia's Excel Coal, a large producer of the metallurgical variety. Australia's low-cost mines kept Peabody profitable, with about 75 percent of the coal being exported to Asia. China remained predominantly coal-powered and, although its proven reserves were vast, so was its energy demand. In 2009, China passed the US as the world's biggest consumer of energy, becoming the largest coal importer. Its coal consumption had doubled over the previous decade to 3.6 billion tons a year (about half the global supply). This, of course, meant that China's CO_2 emissions also soared, to 8,000 kilotons annually by 2009, more than any other nation on earth.

But what spelled doom for the world's environmental health meant prosperity for Peabody. Boyce proclaimed "a global super-cycle for coal" that could last between two and four decades. Meanwhile, Palmer, Peabody's disinformation wizard, rejoiced that the world had entered the first stages "of global hyper growth in energy demand," and that China with its massive coal reserves was poised to be the "new Middle East." The company enlisted outside promotional help from Dr. Frank Clemente, director of Penn State University's Environmental Policy Center, who wrote an article for *American Coal Magazine* based on his research "findings", celebrating the key role coal played in China's economic transformation as well how it would lift other areas of the globe, such as Africa and Southeast Asia, out of "energy poverty." Clemente saw only one dark cloud in coal's future, bemoaning how "climate change is strengthening its unrelenting grip on discourse and making energy poverty an afterthought."

Energy poverty . . . this was Peabody's clever tactic to shift the public conversation away from climate change, with its "unrelenting grip on discourse." Instead of a dirty symbol of environmental pollution, the coal giant would refashion itself as a liberator of the world's poor, promoting cheap coal subsidized by export credits as a way out of the darkness for millions in Africa and Asia who lacked electricity. In September 2010, Boyce announced the "Peabody Plan to Eliminate Energy Poverty and Inequality" in a keynote address given at the 21st World Energy Congress in Montreal. "The greatest crisis we confront in the 21st century is not a future environmental crisis predicted by computer models [climate change]," the Peabody CEO said, "but a human crisis today that is fully within our power to solve. For too long, too many have been focused on the wrong end game." Some 3.6 billion people couldn't readily access electricity, including 1.2 billion children. Thus coal was "the only sustainable fuel with the scale to meet the primary energy needs of the world's rising populations." Boyce later shamelessly cited President Johnson's war on poverty in the 1960s and the UN's anti-poverty Millennium Goals campaign as the models for the Peabody crusade, which, despite Boyce's rhetoric, was nothing more than an effort to further enrich the energy company while muting the clamor over climate change.

The World Coal Association, then headed up by Fred Palmer, made sure that Peabody's "energy poverty" message was circulated around the globe. Burson-Marsteller—a Washington PR firm with a sordid history of working for such notorious clients as the tobacco industry, Union Carbide after the Bhopal disaster, and the Blackwater security firm after the slaughter of Iraqi civilians—was enlisted to put together a marketing campaign. As Kert Davies, director of the Climate Investigations Center, put it, "Burson-Marsteller has spent decades working for some of the world's worst perpetrators of human rights and environmental abuses. So Burson-Marsteller is well suited to help Peabody push dirty coal to the world's poorest people, at a time when everyone from the World Bank to the UN are warning us climate change will hit the poor first, and hardest."

At home, Peabody opposed any regulation that would make its utility clients invest in technological upgrades. But in 2011, appearing before Congress, Palmer had the nerve to praise China for its "unprecedented investment in clean coal technologies." Boyce had good reason to praise the Chinese government. Peabody became the only non-Chinese participant in GreenGen, a power plant and carbon research center in Tianjin. Soon the company entered into partnership with the Shenhua Group, China's biggest state coal company and largest in the world, which made Peabody's thermal coal the country's priority supplier. This was followed by an agreement with China's fourth largest state-owned mining operation for development of a coal mine in Xingjiang that held 40 percent of the nation's reserves.

A founding member of the US Energy Cooperation Program between America and China, Peabody also was brought in by Beijing to help convert coal to synthetic fuel in remote areas. The largest fossil fuel development project on the planet—involving as many as 40 coal conversion plants—was touted as a way to help clean the air in China's heavily-polluted big cities. However, energy experts at Duke and Stanford responded that the massive project would also result in rapacious water use in regions already suffering from drought—and emit between 36 and 82 percent more greenhouse gases than burning pure coal to produce electricity. If all 40 syn-fuel plants were built, according to *Inside Climate News*, we would see 110 billion additional tons of CO_2 released over the next four decades.

Meanwhile, in India—where coal imports were growing more rapidly than anywhere else on the planet—Peabody entered into a joint venture with Coal India (which accounted for some 80 percent of that country's production.) Peabody also expanded its presence in Indonesia, itself the world's leading supplier of seaborne thermal coal, with prime focus on China and India. Much of Indonesia's mining operation occurred in the tropical rainforest, crucial to maintaining the earth's carbon balance.

Foreign coal mines, only 2 percent of Peabody's business in 2003, grew to 40 percent. Peabody opened marketing and trading offices in New Delhi and Singapore, and extended its political influence through-out the world. With a push from Peabody, Australia repealed its carbon tax and emissions trading plans. Boyce praised the country's govern-ment for demonstrating "a lesson in leadership for the modern world.".

In 2014, continuing its energy poverty propaganda barrage, Peabody unveiled its Advanced Energy for Life campaign. Boyce set the stage by writing in the company's corporate social responsibility report: "Peabody believes energy poverty is a human tragedy and a global environmental crisis." The Peabody PR blitz received support from a key player, Bill Gates, founder of Microsoft. The Gates Foundation possessed, at the time, $1.6 million in Peabody holdings. In June 2014, Bill Gates posted two videos about energy poverty on his blog, writing: "Many developing countries are turning to coal and other low-cost fossil fuels to generate the electricity they need for powering homes, industry, and agriculture. Some people in rich countries are telling them to cut back on fossil fuels. I understand the concern. But even as we push to get serious about con-fronting climate change, we should not try to solve the problem on the backs of the poor." Not long after Gates chimed in, Tillerson of Exxon-Mobil began bemoaning the terrible problem of "energy poverty" that condemned people to a life of deprivation if we cut back on fossil fuels.

The poor, Gates and Tillerson failed to mention, will be the ones to suffer by far the most from increasing extremes of weather. Nor did they bring up satellite photos showing the "Asian Brown Cloud" making its way across the Pacific to the US from coal plants with few pollution con-trols. Thanks in large part to Peabody, more than 1,800 new coal plants hit the drawing board in China, India, Vietnam, and Indonesia. "As the Western world cracked down on emissions, [Peabody] targeted emerging economies," as *Climate Change News* put it. The truth is, according to a report by the Australia Institute, solar and wind energy were projected to be cheaper than coal in both China and India within the next decade. But Peabody and its corporate friends kept the focus on "cheap coal" as the savior of the world's poor. Another friend was the World Bank, which, after saying in 2013 that it was ending support for more coal-burning power plants across Asia, ended up funding more than 40 coal projects through its private-sector branch.

A revealing 2015 article in *The Guardian* pulled back the curtain on the Peabody ad campaign. In March of that year, the company posted a three-minute video promoting its Advanced Energy for Life initiative. The video featured a beautiful young woman, who told her moving story over a soundtrack of traditional Chinese music. As a girl, she said, she had to study by candlelight until a coal-fired power plant came to her impoverished village. "Coal . . . is a major force in eliminating fuel poverty in China, but more importantly it's a critical driving force for the phenomenal economic growth China has experienced," she continued. "Eliminating energy poverty will obviously lead to . . . changing the person's perspective to life."

The young lady's name was Linda Jing. And, as it turned out, she was a familiar personality in the world of corporate propaganda. In 2006, she'd been the "face of diversity" for her previous employer, General Motors. Now she worked as an executive at Monsanto, although this wasn't mentioned in the video. Jing told the *Guardian* that she'd not been paid by Peabody. But in May 2015, the company would announce a deal in the Datong mining area where her two older brothers were employed, and where Jing's father worked before being appointed to a senior role by then-Premier Deng Xiaoping.

Peabody's Greg Boyce has his own connection to Monsanto, joining the agribusiness giant's board in 2013, where, ironically, he participates on its sustainability and corporate responsibility committee. Located not far from Peabody's downtown St. Louis headquarters, Monsanto is notorious for its efforts to control the global GMO-seed and herbicide market. But its business strategy also includes ways to capitalize on the coming impacts of climate change.

Back in the late 1990s—when Monsanto launched a new global water business, starting in India and Mexico—executive Robert Farley stated, "What you are seeing is not just a consolidation of seed companies [Monsanto had recently acquired Cargill's seed division for $1.4 billion and DeKalb's for $2.3 billion], it's really a consolidation of the entire food chain. Since water is as central to food production as seed is, and without water life is not possible, Monsanto is now trying to establish its control over water." Monsanto saw depletion of water resources not as a looming global catastrophe, but as a business opportunity. "When we look at the world through the lens of sustainability, we are in a position to see current and foresee impending-resource market trends and imbalances that create market needs," Farley said. Monsanto estimated providing safe water as a several-billion-dollar market.

When the worst drought in half-a-century eliminated half of the US corn crop in 2012, Monsanto looked to reap another boon. That year federal regulators approved Droughtgard, the company's new hybrid corn variety with the first genetically-modified trait designed to withstand a long dry spell in specific locales like Kansas, Texas, and South Dakota. Droughtgard went into the final stages of field testing, with some 250 growers in the Western plains planting about 10,000 acres of the seed. A report by the Union of Concerned Scientists revealed that "despite many years of research and millions of dollars in development costs, Drought-Gard doesn't outperform the non-engineered alternatives."

The same year Boyce took a seat on its board, Monsanto snapped up a San Francisco company, Climate Corporation, which looks at weather data and sells farmers "Total Weather Insurance," enabling them to lock in profits despite drought or heavy rainfall. "As we all know, the weather is becoming more extreme," said Greg Smith, Climate Corp's CEO. "We found that we had kindred spirits with the folks at Monsanto; the data science that we have developed can be applied to improve seed production immensely." Those kindred spirits paid $1.1 billion for the favor. A few months prior to the deal, the US Department of Agriculture's Risk Management Agency had authorized Climate Corporation to administer all federal crop insurance policies for 2014—a potential $20 billion market-of-the-future, Monsanto anticipates.

All the while, wearing his other hat at Peabody, Boyce and his partners worked to undermine all efforts to slow carbon output. While Peabody aggravated the climate change crisis, Monsanto cashed in on it. "Peabody is perhaps the staunchest opponent of stringent regulations to cap greenhouse gas emissions," as *USA Today* said in 2008. Fred Palmer oversaw spending almost $30 million on federal lobbying efforts in this regard between 2006 and 2010. CEO Boyce, for his part, attacked the findings of the Intergovernmental Panel on Climate Change as "tainted by flaws . . . [including] multiple instances of errors, manipulated data, and gaps in information [that] make the IPCC's conclusions unreliable."

Long after climate change became not just scientific consensus, but a daily nightmare for millions around the globe, Boyce was still dishing out the same old disinformation: "Our view is that the globe's climate has been changing since the globe was formed. Levels of CO_2 have risen in the atmosphere, and we have been a strong advocate for technology advances to reduce CO_2 in the atmosphere, particularly from the use of coal." As late as 2014, Peabody was hiring Craig Idso to produce a study titled *"The Positive Externalities of Carbon Dioxide."* A Peabody letter published by *DeSmogBlog* in 2015 still found the company calling CO_2 a "benign gas that is responsible for all life."

That same year, Boyce brought home his energy poverty message to America, declaring in another speech: "High electricity costs put pressure on families, forcing what are too frequently becoming painful sacrifices. No parent should ever have to make the terrible choice of putting food on the table, buying medicine or paying for power, yet these are very real issues for tens of millions of Americans."

Late in 2014, Peabody hired a seemingly strange bedfellow, Laurence Tribe, to make its legal case against the EPA's Clean Power Plan that would force utilities nationwide to cut their emissions. Tribe once taught constitutional law to President Obama, who had served as his principal research assistant while attending Harvard Law School. Tribe had also represented Al Gore before the Supreme Court in the disputed 2000 presidential election. More recently, Tribe had described lawsuits against fossil fuel companies by villagers in Alaska and coastal inhabitants of Louisiana (where sea level rise is already having powerful impacts) as "a profoundly dangerous perversion of the judicial process." The courts concurred.

Now, as he appeared on Peabody's behalf before the House Energy and Power Subcommittee, Tribe lambasted the EPA for allegedly usurping state, congressional and federal court prerogatives. "You know, I've cared about the environment ever since I was a kid," Tribe said. "And you know, I taught the first environmental course in this country, and I've won major victories for environmental causes. But I'm committed to doing it within the law. Burning the Constitution should not become part of our national energy policy." The Obama EPA's plan, Tribe later said, was a "breathtaking exercise of power."

Kentucky Senator Mitch McConnell, the Majority Leader and a staunch climate denier who'd been pushing states to refuse to provide the EPA plans for cutting back on their power plant carbon emissions, unsurprisingly praised the "iconic" Harvard scholar's conclusions. David DiMartino, an adviser to the Climate Action Campaign fighting on behalf of the EPA rule, countered that Tribe's testimony was "phony" and added: "I guess we shouldn't be surprised—a wad of coal industry money burning a hole in your pocket can make you do strange things."

Greg Boyce—whose own deep pockets contained a compensation package worth $57.8 million between 2006 and 2013—was so confident of the power of fossil-fuel money (with the solidly Republican Congress and high-priced lawyers it bought), that he crowed to the *Financial Times*, "[The EPA rule is] never going to happen in the near-term."

While fighting the Obama administration in court, Peabody was simultaneously reaping vast rewards from the federal government's largesse. Mining companies have been given access to publicly-owned coal at subsidized rates for decades. By 1980, Peabody became the largest single company holding federal coal leases in the West, some 82,000 acres, or 8 percent of the total land under lease. It's important to note that Exxon and the other leading oil companies, which had once been major holders of coal reserves, dumped their holdings after their own scientists warned that coal's future was dim. But the dire environmental predictions did not stop Peabody from expanding its reserves.

According to a report issued by Greenpeace in 2014, "The Interior Department has leased 2.2 billion tons of federal coal to the mining industry since the beginning of the Obama administration." Peabody depended on the government's program for 68 percent of the 189,500,000 tons of coal it mined in the US in 2014. The coal dug up from federal land and burned by utilities and industry across the country has resulted in enormous carbon pollution. According to a Greenpeace report, titled "Corporate Welfare for Coal," Peabody's federal coal production alone of 129,313,236 tons in 2014 "resulted in 214,530,808 metric tons of carbon dioxide, or the equivalent of the annual emissions from over 45 million passenger vehicles." Add in the totals from Arch Coal and Cloud Peak Energy, the other prime recipients of the current leasing program, and the annual emissions equal that of 109 million passenger vehicles.

Using the government's own mid-range social cost of carbon figures, the federal coal mined by those three companies in 2014 will cause $18.8 billion in damages when burned.

On paper at least, it has not been all roses for the coal giants in recent years. In 2011, Peabody had gone into major debt while buying another Australian company, MacArthur Coal Ltd., for $5.2 billion, in hopes it would offer a quick-and-easy supply route to China. The company's greatest challenge, Boyce admitted, was finding export routes from the federal leasing bonanza at the Powder River Basin to the Pacific

Northwest and on to China. Peabody announced a deal in 2011 to eventually ship millions of tons of coal from a Gateway Pacific coal export terminal proposed near Bellingham, Washington. But this never materialized. In May 2016, the US Army Corps of Engineers rejected the permit application, bowing to the Lummi Nation's assertion of treaty-protected fishing rights.

So Peabody, along with other coal companies faced with increasing competition from natural gas and the likelihood of much stronger emission controls, was looking at what it called an "unprecedented industry downturn." Since 2011, the overall market value of the US coal sector has nose-dived from over $70 billion to barely $6 billion. In recent years, over 250 coal mines closed. Fifty American coal producers, representing one-quarter of the nation's production, went belly-up. More than 60 percent of the aging coal-fired plants in the US will either have to shut down or be replaced, but no new ones are currently being contemplated. During the second half of 2014, Chinese coal imports also tumbled by more than 20 percent. And thousands of former workers engaged in protests outside Peabody's St. Louis headquarters, complaining they had been cheated out of their health care and retirement benefits after the company spun off its Appalachian mines to Patriot Coal, which then declared bankruptcy and laid them off.

As Peabody reported more than $2 billion in net losses since 2011, while its stock value plunged 93 percent and investors lost over $16 billion, Greg Boyce took a widely-publicized pay cut of ten percent. That meant his base salary would be reduced by $123,000—less than a percentage point of the nearly $11 million he was paid in 2014. The next year, though the company's losses continued to mount, Peabody nonetheless managed to generate $5.6 billion in revenue.

In 2015, Boyce, turning 60, announced he would step down as Peabody's CEO/Chairman at the end of the year, although he would remain on the boards of Monsanto, Marathon Oil, and Newmont Mining Corporation, as well as the US-China Business Council. As he prepared to retire, Greenpeace took note of his terrible legacy:

"Boyce leaves behind a record of environmental destruction, injustice to communities and workers, laughably incorrect predictions, and dismal results for Peabody investors—all while pocketing tens of millions of dollars. As *Rolling Stone* put it, 'There's no better example of how capitalism profits from overheating the planet than Boyce.'"

Also departing Peabody was Fred Palmer, who, after working his way up to be Boyce's special advisor, jumped ship to the Phoenix-based PR firm Total Spectrum. There, Palmer's M.O. would remain much the same, but on behalf of energy interests in his native Arizona.

Boyce's successor, the Australian-born Glenn Kellow, took over the reins just as several other St. Louis-based coal companies—and then Peabody itself—declared Chapter 11 bankruptcy, citing "a dramatic drop in the price of metallurgical coal, weakness in the Chinese economy, overproduction of domestic shale gas and ongoing regulatory challenges." As Peabody "restructured" its debts and laid off 15 percent of its workforce, largely in the US, the company made clear that its financial troubles would not affect its expansion plans overseas, where hundreds of coal plants were still on its drawing boards.

As Peabody and its fellow St. Louis coal companies filed their bankruptcy papers, the documents revealed interesting facts about their climate-change denial campaigns. It turned out, for instance, that the chief truth-spinner for Alpha Natural Resources—the fourth-largest US coal company—was Washington attorney Chris Horner, a Fox News regular who claimed the earth was cooling and is a senior fellow at the Competitive Enterprise Institute. In 2009, Horner had instigated what he called "Climategate"—wrongly accusing scientists at the University of East Anglia of "falsifying results, collaborating to subvert and violate the laws" in a series of emails. Six official investigations cleared the scientists of any wrongdoing, but Horner continued to flood climate researchers at major American universities with records requests, seeking to distract their efforts.

Horner was also a recipient of Peabody funds, according to its bankruptcy court filings, which also revealed payouts to at least two dozen other groups that disseminated false information on climate change. In fact, the list of Peabody cash recipients was "the broadest list I have seen of one company funding so many nodes in the denial machine," according to Kert Davies of the Climate Investigations Center. Among Peabody's benefactors were the American Legislative Exchange Council (the corporate lobbyists who, among other things, have sought to make homeowners pay financial penalties for installing solar panels), discredited Harvard scientist Willie Soon, and the evil PR genius Richard Berman.

After the 2016 election, the incoming Trump administration named Horner to its EPA transition team.

Peabody's declared debt of $6.3 billion is currently being restructured in the St. Louis courts, where the company is seeking not only to reduce its debt level, but to improve cash flow and "position the company for long-term success, while continuing to operate under the protection of the court process." In short, to gain relief from its obligations to shareholders, creditors, employees, retirees, and the general public.

"There is a deeper story here—who is profiting off this bankruptcy and who is really bearing the brunt of the burden," says Caitlin Lee, of the nonprofit group, Missourians Organizing for Reform and Empowerment. Lee's introduction to Peabody came as a recent undergraduate at Washington University of St. Louis, when hundreds of students occupied the administration building to protest the school's ties to the company.

In particular, the protest focused on Greg Boyce, named in 2009 to the university's board of trustees. Shortly before his appointment, Peabody had teamed with Arch Coal and the St. Louis utility Ameren to donate $5 million toward funding a Consortium for Clean Coal Utilization at the school. This would coincide with tax breaks provided to Peabody by the city, on $61 million worth of purchases and improvements to its downtown headquarters—$2 million of which was taken out of the budget for public schools.

"As climate change escalates and as it becomes a major issue for young people in my generation," said Caroline Burney of Washington University Students Against Peabody at the time of the 2014 action, "we've realized that we can't really wait and that we need to escalate our tactics and show the administration that students really care about this issue."

Seven university students were arrested while trying to speak to the school's board of trustees about its ties to Boyce. Then, in 2015, while a few students handed out flyers outside the annual meeting where Boyce was up for another term, "they brought in riot gear and were ready with barricades" to prevent another occupation, according to Caitlin Lee. Peabody happens to be a major donor to the St. Louis Police Foundation, established in 2007 to provide "a safety net to fund critical needs."

Boyce is still on the university's board of trustees. Meanwhile, the university's consortium, which still lists Peabody as a lead sponsor, continues "to foster the utilization of coal as a safe and affordable source of energy, and as a chemical feedstock, with minimal impact on the environment."

So what of the Peabody bankruptcy? The deeper story, as Lee describes it, revolves around how the company reconstitutes itself. The bankruptcy judge, Barry Schermer, is a prominent professor at Washington University. A number of the school's graduates or current teachers are representing various parties involved. Profiteers including "vulture hedge fund" operator and Republican donor Paul Singer are involved, too. And Peabody's creditors include Citibank and Franklin Resources, which will be determining the company's liabilities.

How much of the bill will the taxpayers end up footing? Will Peabody be able to shift its massive environmental clean-up costs to taxpayers? The price tag alone for reclaiming the land strip-mined by Peabody has been estimated at $1.4 billion. It should be noted that in North Dakota and Wyoming, where 450 square miles of land were shattered by coal mining, only 10 percent has been reclaimed.

Coal companies are, in principle, required to post bonds to pay for this work. But in Missouri and other conservative Republican states, as

Carl Pope, former Executive Director of the Sierra Club, has written, "Those with creatively puffed up balance sheets were allowed by states to 'self bond' and guarantee their own obligations." These sites aren't just ugly—they are perpetual polluters, slowly but surely sending acidic and other toxic wastes into nearby streams and into groundwater.

Additionally, Pope writes, "Coal companies hold in trust the funds that pay their workers' pensions and health care. But over the past few years companies like Peabody Coal divided themselves into bits and pieces. Certain bits kept the parts of the coal business Peabody thought were money makers. Other bits (with sexy names like Patriot Coal) got assigned mines that were no longer profitable, along with bundled liabilities like retiree pensions and health care. When Patriot Coal, predictably, filed for bankruptcy, it stuck the taxpayers and the workers with the obligations."

Greg Boyce must be laughing all the way to the bank. The ascension of Donald Trump saw an immediate rally in coal prices and thus Peabody stock. Its bondholders stated in a court filing that the president-elect's intentions "to revitalize the coal industry and to roll back the regulations" would be a boon. Indeed, Peabody bondholders alleged, metallurgical coal prices had climbed 235 percent and thermal coal prices were up 77 percent since the bankruptcy filing.

Peabody has moved ahead seeking approval from the new administration to expand its coal mine on Navajo and Hopi land in northern Arizona. Since 1987, Peabody has paid about $50 million a year to the two tribes for permission. Now several Navajo NGO's are going up against their leadership to join the Sierra Club in a legal fight to stop the expansion, saying that burial grounds and pre-Columbian ruins would be destroyed in the process. The rub is, the mine happens to fuel a power plant majority-owned by the permitting agency, the US Bureau of Reclamation. In Trump world, that's not a good sign.

Some 320 miles south of Chicago, just east of the southern Illinois town of Harrisburg, is an area known as Rocky Branch. It's close to where

Francis Peabody drilled his first mine in 1895, and for many years was a thriving rural community. Today a mine that Peabody opened in 2014 threatens to wipe out more than a thousand acres of farm and forest land on the edge of the Shawnee National Forest. Its blasting operations come as close as 300 feet to the homes of residents who've refused to sell their property to Peabody.

Georgia de la Garza, founder of the Shawnee Hills and Hollers grass-roots group, stops her car and points to the nearby blackened hill. "The coal here is very close to the surface," she says. "This is actually light production right now, but when it comes back up, the blasting is so loud and with lights that shine 24/7. A lot of people around here wear earplugs. It's terrorism I feel, nothing short of."

Many families have left. Others chose to profit from leasing mineral rights to the corporation. Still others, knowing full well the impact of airborne coal dust and polluted waterways at Peabody's now-abandoned Cottage Grove strip mine a mile away—which left only a cemetery behind—decided to fight. In March 2014, residents of Rocky Branch staged a three-hour-long road blockade. That summer, one of them—a decorated Vietnam War veteran, Glenn Kellen—got arrested as he tried to move his protest sign closer to the public road.

The trucks kept coming. Delivering a speech at the time in Australia, Greg Boyce said, "Coal always wins."

De la Garza turns around and heads across old Highway 13—bequeathed to Peabody by the state for a nominal fee—to where Muddy Road runs behind Southeastern Illinois College. The company has donated a building to the college, where several coal companies train engineers. As a coal truck passes slowly by, De la Garza stops again by a sign: Coal Refuse Disposal Area #2. It's a 160-acre impoundment structure with an orange residue trickling along the surface of a waterway.

"They spray an adhesive constantly, but you can see it erodes and cracks," de la Garza says. "The Henry aquifer sits on the far end of this. It feeds over 25,000 people their drinking water. We're in a flood zone so a lot of the water that's contaminated with heavy metal escapes. On the other side the impoundment is surrounded by agricultural fields. There are so many sites like this in southern Illinois, it just boggles my mind."

She drives by the abandoned mines known as Cottage Grove and the Wildcat Hills Underground Mine, both owned by Peabody, then down the road to a large ongoing operation owned by Robert Murray. Headquartered in Ohio, Murray Energy is the biggest privately-owned coal outfit in the US, according to its web site, "producing approximately 65 million tons of high quality bituminous coal each year, and employing over 7,500 people in six states."

Murray laid off 1,800 workers, just over 20 percent of its company, in 2015 –a problem that the owner explains like this: "Americans must understand that the closure of 411 coal-fired power plants, over 100,000 megawatts, by our current [Obama] administration jeopardizes the reliability of our electric power grid and the low cost of electricity. The destruction of our industry will hurt poor families the worst and make the manufacturing of American products less competitive in the world marketplace. Low cost electricity, a staple of our lives, is threatened."

Murray has for years fought federal efforts to exert any semblance of control over his industry—from acid rain and mercury emissions to mountaintop removal and greenhouse gases. "The earth has actually cooled over the last 15 years," he claims to believe. In December 2015, at a meeting of climate change deniers in Texas, Murray praised US Representative Lamar Smith for investigating prominent climate scientists and environmental officials, including issuing a subpoena for the head of the National Oceanographic and Atmospheric Administration. The Union of Concerned Scientists and American Meteorology Association need to be scrutinized by Congress as well, Murray said, because those "crony capitalists [are] making a fortune off you the taxpayers." Murray went on to organize a fundraiser for Ted Cruz, after the presidential candidate held a hearing challenging climate research that he called "Data or Dogma?" But when Cruz's campaign fizzled out, Murray held an invitation-only fundraiser for Donald Trump in West Virginia coal country. According to de la Garza, "Our coal is easy to get to, so Murray is pulling a lot of employees from West Virginia to here."

The Illinois highway on which Peabody, Murray and Alliance run their coal trucks is known locally as "death road." The coal "goes down to the Ohio River, where they fill the barges up and then join the Mississippi. Our biggest contractors to burn Illinois coal are China and Indonesia. The coal barons ship it overseas and we're left with people losing their homes and an extraordinarily high rate of cancer."

She was born and raised here, Cherokee on both sides of the family, her ancestors having settled in southern Illinois during the infamous "Trail of Tears" migration. Her grandfather had been a teenage coal miner who eventually died from black lung disease. Her father built the local malls and ran a bait shop and oversaw the Pro Bass fishing tournaments. "We were hardly ever at home, always down in the forests," de la Garza remembers. "We used to ice fish, but you can't do that anymore with climate change. Might be able to sled one day a winter where we used to go for months."

In April 2016, St. Louis and southern Illinois community organizers—including Caitlin Lee and Georgia de la Garza - brought in a group of Native Americans from the Navajo Black Mesa reservation in Arizona, where Peabody has two strip mines. They joined together in a rush-hour street blockade outside the company's downtown headquarters. They called for a "Just Transition Fund" for communities harmed by Peabody's operations, as part of the bankruptcy settlement.

On her Facebook page, de la Garza posted a video featuring a teenager from Black Mesa. He read aloud a prose-poem that he'd written, addressed directly to Peabody's CEO. It concluded:

"Dear Mr. Boyce—Greg—someone tells me you are a nice guy. I believe that, that you are just one person and you don't represent all the gross injustices of American institutions. I don't believe you're evil. I believe you're a coward.

"How easy it must be for you to forget the past, to ignore the present, to shut out the world you are making - and let it burn."

Chapter 8

The Kochs In Kansas— Land of the Fee and Home of the Crave

Towering above the prairie on the northeast outskirts of Wichita, the global headquarters of Koch Industries is impossible to take in all at once. The buildings that loom over its corporate campus are designated North, South, East and West. The largest is a hulking, box-shaped structure that features a tall American flag in front of opaque smoked-glass windows. Each building is surrounded by security fences, made somewhat less foreboding by rows of Kansas's emblematic sunflowers.

With the addition of Building H in 2016, the campus sprawls across more than 1.2 million square feet and hosts some 3,600 employees. The new building is an anomaly, both for Koch Industries and for Wichita. It was designed to meet the EPA's Energy Star requirements for efficiency—white roof and special window coatings to minimize the impact of the hot summer sun, all-LED interior lights, even a 20-million-gallon retention pond for irrigation of the surrounding landscape. At the groundbreaking ceremony, hundreds of employees wore matching orange shirts with "Creative Construction" emblazoned across the back. "A fundamental part of our belief," Charles

Koch told the crowd, "is to provide products and services that make people's lives better."

The 80-year-old Koch resides not far away, close to the Wichita Country Club, inside a large, leafy estate with a tennis court and a fountain shaped like an ancient temple with its eight cruciform symbols and four equilateral triangles. In close proximity are the Koch Community Plaza, the Koch Scouting Center, and the Koch Orangutan and Chimpanzee Habitat at the Sedgwick County Zoo, where the company holds its annual picnic. Local children take swimming lessons at the Koch Aquatic Center and visit the Koch Habitat Hall inside the Great Plains Nature Center. Koch Industries fosters after-school programs and offers financial support to Big Brothers Big Sisters.

At Wichita State University, where the basketball arena is named after the Kochs, the foundation started by Charles's parents recently donated $3.75 million toward a design space for student inventors. "A sexy project with young, motivated people," Charles's wife Liz told the local newspaper. "We are at the place where we need to pass the baton around here." She added that she has two grandchildren, "but I want ten more. . . . The smell of diapers is like perfume to me."

And what of the family's children? Their son Chase, now in his late 30s, joined the family enterprise after college. He still lives in Wichita. After promoting Chase to the presidency of Koch Fertilizer, his father commented without irony, "At every step, he's done it on his own." Chase holds special events for the children of wealthy contributors at the Koch brothers' yearly donor seminars.

Charles's daughter Elizabeth, however, isn't cut from the same mold as her brother. A Princeton graduate with a master's degree in creative writing from Syracuse, she owns a small publishing house in Brooklyn called Catapult. In Jane Mayer's *Dark Money*, Elizabeth is described as "chasing" her father around the house, seeking to impress him with her interest in economics and confiding to her journal about "staring down that dark well of nothing you do will ever be good enough you privileged waste of flesh." The first book Catapult published was titled *Cries For*

Help, Various. (Elizabeth Koch did not respond to an interview request for this book.)

Her father Charles is himself a published author, specializing in boosterish type of business advice books that might be subtitled "Things Go Better With Koch." His latest book, *Good Profit*, is follow-up to his earlier *The Science of Success.* According to his wife, Charles still goes to work every day at 7 AM, "obsessed about getting things done, and making life better for everybody." She also told the interviewer from the Wichita paper, "People have made him into this monstrous person that I've never seen."

There is a photo online of Charles Koch laughing. He is wearing a sweatshirt emblazoned, "Billionaire Boys Club."

As recently as 2015, Charles Koch stated publicly that yes, the climate has "been warming some," but "it's not certain" that humans are to blame. In a May 2016 interview with ABC News, he elaborated: "I believe the evidence is overwhelming that it's changing in a mild and manageable way. These policies that are being introduced in the United States, as a matter of fact, under their own models would have virtually zero impact on the future temperature or other aspects of the climate. And in fact I think they make matters worse, because they get people going after the subsidies rather than innovating."

Koch also trotted out the same theme featured in Peabody's "Energy Poverty" campaign and by Exxon's Tillerson, warning about pollution-control policies that "are making people's lives worse. They're raising the cost of energy for no benefit and guess who suffers the most—the poorest people use three times the energy as a percentage of income than the average American."

As with other energy moguls, Charles's arguments against climate-change action evolved over time. In the past, Charles voiced doubts about the scientific consensus and even suggested that global warming is good for us. A column he penned headlined "Blowing Smoke" for the company's in-house newsletter mused, "Since we can't control Mother Nature, let's figure out how to get along with her changes."

Charles Koch's younger brother and corporate partner, David, has gone even further in praising the possibilities of climate change, in a kind of "how I learned to stop worrying and love the Bomb" kind of way. "The Earth will be able to support enormously more people because a far greater land area will be available to produce food," he has said. At the Smithsonian's David H. Koch Hall of Human Origins (which opened in Washington in 2010), the message of an interactive game was that in the event climate became intolerable, we would simply build "underground cities" and develop either "short, compact bodies" or "curved spines" so that "moving around in tight spaces will be no problem." Really. And this is in a national museum that celebrates human knowledge.

David Koch is said to be the wealthiest man in New York, a major benefactor of Lincoln Center and the Metropolitan Museum of Art among other institutions. The dinosaur wing of the American Museum of Natural History is named after David, who gave $20 million to fund it. ("I was gaga about dinosaurs as a kid," he has said). Here too, Koch has used the museum as a platform to spread his "what, me worry?" attitude about climate change. The dinosaur exhibit touted the theory that mankind has always evolved in response to climate change. In 2016, David stepped down from the Natural History museum's board, saying he no longer had time to attend the meetings. Or perhaps he realized that a museum that showcased extinct creatures was not the best advertisement for the continued pumping of CO_2 into the atmosphere.

The two Koch brothers' wealth was estimated in 2014 at more than $50 billion apiece. Their companies are an outgrowth of the business their father Fred started in Kansas in 1925, where he and a business partner developed a superior process for refining oil but were forced out of the American market by a series of patent infringement lawsuits from competitors. This led Fred Koch to find his fortune in Stalin's industrializing Soviet Union and then by building refineries in Hitler's Germany. From this tainted corporate past, Koch Industries has grown to employ more than 100,000 people in over 60 countries and brings in some $115 billion a year in revenue. Koch Industries spans a variety of commercial pursuits, from refining to ranching to fertilizers and forest products. The Kochs own more than 4,000 miles of pipelines, and are the planet's largest exporter of oil from the Canadian Tar Sands. Theirs is the second largest private company in the US.

In 2014, a study of corporate polluters published by the Political Economy Research Institute at University of Massachusetts Amherst listed Koch Industries' annual greenhouse emissions at 28 million metric tons—22nd on its list and ahead of such oil companies as Chevron, Royal Dutch Shell and Valero. That's as much CO_2 as comes from about five million automobiles.

Perhaps this is why the brothers have poured a vast fortune over the years into the political effort to block climate-change reforms—outspending ExxonMobil by three-to-one and lining the coffers of GOP climate deniers on Capitol Hill. Two decades ago, Koch became the first major corporation to enlist a tax-exempt nonprofit as a front group, described as "a cutout to secretly influence elections in a threatening way" by Charles Lewis of the Center for Public Integrity. Recently, Koch has used Donors Trust in Alexandria, Virginia, as a conduit "behind which fingerprints disappeared from the cash," according to *Dark Money* author Jane Mayer. Between 2005 and 2011, Donors Trust received nearly $8 million in grants from Charles Koch's foundations—cash that was then funneled to groups like the Americans for Prosperity Foundation (AFP), conveniently chaired by David Koch. (To get a tax deduction, foundations can't donate to themselves.) In addition to helping organize the Tea Party's "grassroots" rebellion against the Obama administration (and the Republican establishment), the AFP assumed leadership of the brothers' national effort to stop any action on climate change. The AFP now boasts offices in 35 states and has even established a Grassroots Leadership Academy to hand out certificates to "community mobilizers."

At the federal level, the Koch Industries PAC was the largest contributor to members of the House Energy and Commerce Committee. By the beginning of the 2011 legislative session, the Kochs' "No Climate Tax" pledge had been signed by 156 members of Congress. Mitch McConnell, the Senate Majority Leader, hired a former Koch lobbyist as his policy chief before initiating a full-scale attack on the EPA, during which he pushed governors to refuse compliance with the environmental agency's new restrictions on CO_2 emissions.

Koch money also helps fuel the pro-fossil fuel lobbying of groups like the State Policy Network, a nationwide coalition of conservative think-tanks, and the American Legislative Exchange Council (ALEC), which fight renewable energy legislation at the state and local level. In 2013 alone, these Koch-funded groups targeted some 70 pieces of clean energy legislation.

In 2016, the Koch brothers didn't back a presidential campaign and claimed to be focusing their largesse on a dozen Senate races. But in 38 states, they ended up deploying some 1,600 paid staff to push policy agendas and back Republican contenders. The Trump transition teams were dominated by Koch political hacks. And they will surely profit handsomely from the new administration's plan to move ahead with the Keystone XL pipeline. As mentioned, Koch Industries is the world's biggest exporter of oil from the Canadian tar sands.

Trump's choice to head the EPA, Scott Pruitt, will be warmly welcomed by the Kochs as well. Prior to Pruitt's confirmation hearing, every US senator received a letter from a coalition of 23 nonprofit groups backing him. No less than eighteen of these organizations had received generous support from the Koch brothers' many foundations. Eight of them collectively received over $30 million between 2010 and 2014 from American Encore, established as a "social welfare"nonprofit by the Kochs.

Emails released under a court order soon after Pruitt's confirmation reveal that he and his Oklahoma Attorney General's office maintained a close relationship with the advocacy and legal groups funded by the Kochs, including Americans for Prosperity's Oklahoma State Director, John Tidwell. Pruitt was a sought-after speaker at their gatherings, including the 2013 ALEC conference, where he was part of a panel on

"Embracing American Energy Opportunities From Wellheads to Pipelines" and attended a luncheon sponsored by Koch Industries.

Nowhere is the noxious influence of Koch Industries money more strongly felt than in the state it is based, where the corporate giant has taken aggressive measures to block the rise of renewable energy. As the world has known ever since Dorothy got spirited to Oz by a tornado, weather in Kansas can be way beyond breezy. It's said that the wind blows through the state strongly enough to meet its electricity needs more than 90 times over. According to the American Council on Renewable Energy, Kansas "has one of the most promising wind resource potentials in the country." Big companies including Google and Walmart have looked to invest in the market there. Polls show almost 90 percent of voters in this very red state support the development of renewables.

For a while, wind was a winner. In early 2007, a Kansas housewife named Dorothy Barnett heard a presentation by longtime environmental activist Wes Jackson of Kansas's Land Institute, referring to Kansas as "the Saudi Arabia of wind." Jackson's daughter-in-law had just started the Climate & Energy Project, and Barnett got inspired to form a wind energy task force. "We worked across party lines and built coalitions of land owners, county commissioners, and farmers that were never on the same page before," she recalls. Their emphasis wasn't on climate change, but the sound economics of wind energy.

In 2009, Kansas's State Legislature passed the Renewable Portfolio Standard (RPS), signed into law by then-Democratic Governor Mark Parkinson. A RPS, eventually in place in 29 states, mandates that renewables comprise a portion of a state's energy mix. In Kansas, investor-owned utilities were now required to have 20 percent of their power generated from renewables by 2020. Even conservative Republican Sam Brownback, elected governor in November 2010, was initially an enthusiast. As a US Senator, he'd already cosponsored a federal version of the RPS (though it failed to pass).

Kansas quickly soared well ahead of the game. Within two days of the RPS making it through the state legislature, the Siemens corporation announced it would build a $50 million state-of-the-art wind generator in Barnett's hometown of Hutchinson, bringing 400 jobs "in one fell swoop," as she put it. By 2014, wind provided almost 22 percent of the state's electricity generation, third highest in the US Twenty-five wind farms supplied nearly 3,000 megawatts of capacity, with the industry employing close to 12,000 people and farmers receiving between eight and ten million dollars a year from leasing of their land. Especially compared to the volatility of gas prices and increasing federal regulation of coal (a proposed new coal plant in Holcomb, Kansas, had been squelched by a Sierra Club lawsuit), wind energy was low-cost. As one observer put it, "Even people who would never believe in climate change if you paid them wanted wind farms."

Then Koch Industries stepped in. It started with the 2012 election. First came rumors that the Kochs, who were Brownback's leading campaign donors, might not support his bid for reelection if he kept touting wind energy. The governor soon fell in line. Then the Kochs targeted the state legislature. At the time, recalls Moti Rieber - an ordained rabbi who also heads up Kansas Interfaith Action, a nonprofit activist organization with a primary mission of combating climate change—"there were nine moderate senators in the legislature. The Kochs, through their front group Americans for Prosperity and the Kansas Chamber of Commerce—of which the Kochs are the major funders—targeted these senators in the primary and got rid of six of them. The Senate then became a right-wing cesspool. The Kochs remade the government in their image."

That same year, which marked the first election cycle affected by the Supreme Court's *Citizens United* decision, the Kochs' network of now-secret donors also made a major impact at the national level, spending at least $407 million through 17 tax-exempt conservative groups, and easily retaining Republican hold on the House of Representatives. Among the candidates toppled by the Koch machine was Democratic congressmember Tom Perriello of Virginia, who had pushed for a cap-and-trade bill. His victorious opponent focused on only one issue: the threat that climate change action posed to the economy. Shortly after, the Kochs and their network launched a successful national campaign to let expire a federal tax credit for financing renewable projects.

Back in Kansas, the next legislative session saw the first all-out effort to get rid of the pesky RPS. During the first four months of 2012, the Kochs' Americans for Prosperity—a group singlemindedly concerned with the Kochs' own prosperity—spent $383,000 on media ads calling for the statute's repeal. The group flew in climate change-denying scientist Willie Soon from the Harvard Smithsonian Center for Astrophysics to testify at a legislative hearing. Soon had personally received $230,000 in grants from the Charles G. Koch Foundation between 2005–2012. Soon was joined by a similar mouthpiece from the Heartland Institute. The scientific point of view on climate change—that is, the credible one—was in short supply during the Kansas legislative hearings. "A climatologist from the University of Kansas who had been part of the IPCC [Intergovernmental Panel on Climate Change] was given ten minutes out of the two hours of presentations to try to refute the nonsense they were spewing," says Dorothy Barnett. This was standard operating procedure during the hearings in Topeka. The legislature also placed limits on how much time advocacy groups could testify, forcing activists to blend their presentations. According to Rabbi Rieber, "We had to co-write our testimony. All of us had talking points in there—the renewable energy program helps rural districts, it's good for the environment, and so on. We brought in a salt-of-the-earth Kansan to deliver it."

In the end, Governor Brownback found the Kochs and their money more persuasive than he did the protectors of his state's environment. But, as he moved to phase out the RPS program, Brownback encountered pushback from 12 Kansas companies with a stake in renewables, and that year, the repeal bill didn't get out of committee.

Nonetheless, Rieber, who helped organize a new coalition called Kansans for Clean Energy, knew it would be an uphill fight to keep the mandate. "We were told that Charles Koch's top priority was still gutting the RPS," Rieber says. "Well, in 2014 they spent $500,000 against this ragtag band of misfits—and they lost." While the Kansas Senate voted for repeal, the House narrowly elected to keep the RPS in place. The following year, anti-RPS legislation pushed by the Kochs also failed in Colorado, New Mexico and New Hampshire. Even in Texas, a large coalition of rural conservatives joining with local businesses managed to keep the standard.

And yet, the Kochs still kept coming. "They have unlimited amounts of money and time; we're basically volunteers, so it's not a fair fight," says Rieber. "The Kochs don't do anything themselves, they have capos. They don't meet with the legislators that they buy off either, they use intermediaries. The Kochs are the major funders of the Kansas Chamber of Commerce. The AFP and the Kansas Policy Institute—think tanks that the Kochs seed—promoted the falsehood that the RPS had raised people's electric bills. They beat it into people's heads eight hours a day on radio ads—all paid for by the Kochs."

So finally, in May 2015, the Kansas Legislature voted to replace the mandatory RPS with a voluntary commitment. It happened behind closed doors where no clean energy advocates were allowed, in meetings between the AFP, Kansas Chamber, legislators, and representatives from the wind industry—which had been threatened with paying a 4.33 percent excise tax on their production. AFP state director Jeff Glendening called the deal, which was spearheaded by Koch lobbyist Mike Morgan, "a good compromise that promotes the market." Governor Brownback said, "This further solidifies and stabilizes the policy environment so that investment can continue in Kansas."

But the Sierra Club saw it for what it was—a "backroom deal" that would penalize Kansas's nascent wind industry, giving the neighboring states of Nebraska, Oklahoma, and Iowa a competitive advantage. Iowa, which has far less wind than Kansas, has since surpassed it in generating capacity.

Why was Koch Industries so determined to stop the wind effort? "The only thing I can think of is, it's about market share," says Dorothy

Barnett, today the executive director of the Kansas-based Climate & Energy Project. "In the early days nobody thought wind energy would grow to where it did. Then suddenly with technology upgrades and it becoming more cost-effective, wind became a threat. So I don't think they [the Kochs] are these evil people plotting America's or the world's demise. But I think they have a warped sense of what the future should be and feel they have enough power to influence whatever they want."

President Obama himself took the Kochs to task in the summer of 2015, commenting, "It's one thing if you're consistent in being free market. It's another thing when you're free market until it's solar that's working and people want to buy and suddenly you're not for it anymore. That's a problem." Charles Koch responded that he was "flabbergasted" by the president's comments. "We are not trying to prevent new clean energy businesses from succeeding," he told *Politico*. "Any business that's economical, that can succeed in the marketplace, any form of energy, we're all for. As a matter of fact, we're investing in quite a number of them, ourselves—whether that's ethanol, renewable fuel oil . . . We're investing a tremendous amount in research to make those more efficient and create higher-value products."

Charles apparently saw no contradiction between his response to Obama and what he'd written in a piece posted on Koch Industries' web site: "It takes courage to turn down an investment in a solar energy firm that is guaranteed to be profitable. But only because of its federal subsidies and mandates." Or this, about the wind industry: "Cronyism enables favored companies to reap huge financial rewards, leaving the rest of us—customers and competitors alike—worse off. One obvious example of this involves wind farms. Most cannot turn a profit without the costly subsidies the government provides. Meanwhile, consumers and taxpayers are forced to pay an average of five times more for wind-generated electricity."

Koch likes to pose as a free-market purist. But while attacking renewable energy subsidies, he conveniently leaves unmentioned the sizable government contributions received by the oil and gas industry. In 2015, researchers at the International Monetary Fund released a report describing how the world's governments provide $5.3 trillion in benefits annually to fossil fuel companies. By contrast, according to analysis by a venture capital firm (DBL Investors), between 1994 and 2009, renewable energy technologies averaged only $370 million in annual tax breaks. Nor did Koch say anything about how Congress exempted developers of natural gas from major provisions of the Clean Air Act, Clean Water Act, Safe Drinking Water Act, and at least four other federal laws.

The future of wind power was not the only energy debate that embroiled the Koch-controlled Kansas state capitol. As President Obama escalated his climate change efforts during his second administration, the Kansas Legislature dug in its heels, debating in 2015 whether to comply with the EPA's Clean Power Plan (CPP). This set of regulations requires states to meet goals to reduce their carbon emissions from power plants over 30 percent by 2030. Predictably, Governor Brownback alleged that the CPP violated states' rights and would result in big utility cost increases. So a hearing ensued in the Kansas Legislature chaired by Dennis Hedke, an oil-and-gas geophysicist who built his career as a climate change denier en route to chairing the state's House Energy committee.

Blasting the EPA rule, Hedke claimed, "They have overstepped so many bounds it's just almost unconscionable." Hedke's complaint overlooked the fact that Kansas had been given one of the least stringent state goals, according to the EPA—and, because of the wind industry, was already well on the way to achieving them.

When Rabbi Rieber identified climate change as his reason for testifying, he was told by Hedke, "You have 30 seconds." The rabbi later called it "a kangaroo hearing," writing of how one state senator "implied that a violent rebellion was in store if the government kept it up."

In fact, Rieber wrote in an op-ed for the *Wichita Eagle*, "Kansas is well-situated to take advantage of the opportunities presented by the CPP. Almost 20 percent of our electricity grid is from renewable energy already, and that number can increase. We can export our wind energy to states to the east that have to meet emissions targets but don't have the abundant renewable resources we have. Our substantial capacity for solar energy has only begun to be utilized. And Kansas is consistently among the lowest states in the union when it comes to energy efficiency, which is the lowest hanging fruit of all."

Nonetheless, in October 2015, Kansas joined 26 other states to challenge the EPA rule in federal court, an initiative spearheaded by Scott Pruitt in the neighboring state of Oklahoma. Kansas' new chief deputy attorney general, Jeff Chanay, proved instrumental in the effort, claiming, "There is just no way possible to comply with these implementation dates." In February 2016, the US Supreme Court placed a temporary hold on the requirements of the EPA's CPP, dealing a significant setback to Obama's climate change program—now in grave peril with the new administration.

While the Koch Brothers pursue policies that ensure the planet will be less and less habitable for future generations, they are also trying hard to convince young people they have nothing to worry about. The Kochs are doing to the academic environment what they have been doing for decades to the natural environment. The Charles Koch Foundation sponsors what its vice-president calls a "robust, freedom-advancing network" of close to 5,000 professors at some 400 colleges and universities across the country. They teach courses such as "Free Market Environmentalism," using textbooks like one titled *Economics: Private and Public Choice* that debunk climate change science. "Only idiocy would conclude that mankind's capacity to change the climate is more powerful than the forces of nature," stated George Mason University's Walter Williams, one of the Koch-funded professors. At the University of Colorado, another professor, Steven Hayward, created and starred in the climate-denial documentary, *An Inconvenient Truth . . . or Convenient Fiction?*

In Kansas, the Kochs aimed younger, establishing a Young Entrepreneurs Academy in high schools in the state capital of Topeka. Aimed at low-income school districts, the academies instruct that the poor are being hurt by minimum wage laws and welfare programs. Twenty miles away, at the University of Kansas (KU), Art Hall is the director of a Center for Applied Economics. A former chief economist for Koch Companies Public Sector, Hall's university effort got kicked off by a half-a-million dollars from the Kochs. Among Hall's chief tasks was seeking repeal of the RPS, about which he testified in 2014 before the Kansas Legislature.

Schuyler Kraus, president of the KU branch of the national campaign UnKoch My Campus, filed an Open Records Request looking to uncover more about Hall's ties to the Kochs. The school prepared to comply until the professor filed an injunction at the last minute to prevent disclosure of the documents. He claimed this would constitute a violation of his academic freedom. His legal bills were paid by the Kochs. After a judge upheld Hall and blocked release of his correspondence, in August 2015 the university did make public a limited amount of information confirming that Hall's center was Koch-created.

Mike Pompeo, the US Representative from the Wichita area, is leaving that job to be CIA Director in the Trump administration, without an ounce of national security experience. Pompeo is a Tea Partier and the biggest recipient of Koch campaign funds in Congress. They had previously invested an undisclosed amount in an aerospace company founded by Pompeo. Pompeo signed onto the "No Climate Tax Pledge," the initiative pushed by the Koch-funded Americans for Prosperity front group to resist enacting a climate tax on the big energy producers. In 2013 Pompeo introduced the Natural Gas Pipeline Permitting Reform Act, a bill aimed at speeding up the approval process.

He continued to faithfully pursue a Koch agenda, introducing legislation to scuttle the existing tax credits for the wind industry, saying it should "compete on its own." Some 46 of the bill's 53 recent co-sponsors had received Koch Industries' money. The Kochs managed to get an op-ed into the *New York Times* co-authored through a trade association called Freedom Partners, claiming that the wind tax credit would cost the Treasury some $10.5 billion over the coming decade. (An August 2016 *New York Times* article on the growing use of reputable think thanks as fronts for corporate campaigns acknowledged that the op-ed pages of the *Times* are sometimes used by these think tanks as platforms for their corporate-sponsored ideas.) However, despite the Koch offensive, the budget passed by Congress in December 2015 included a multi-year extension of wind power tax credits—thereby ensuring "stability for 73,000 American wind industry workers across all 50 states and private investors helping [it] to grow," as the wind industry trade association put it. In Kansas, this means the wind industry will be close to generating 4,100 megawatts of capacity by the end of 2017.

So the Kochs and their minions don't always succeed. Even in Kansas, there are signs that the winds are shifting. The state's largest wind farm to date was recently announced by Kansas City-based Trade Wind, which identified the Kansas City Board of Public Utilities and Google as two of the primary recipients of the project's wind energy. Another sign of change: the Koch name is increasingly dirty, even in the company's home state. The Kansas branch of the Nature Conservancy, which has taken millions of dollars from the Koch brothers in the past, recently announced it will not be taking any more Koch cash.

Dorothy Barnett has created a Clean Energy Business Council, gaining support from the state Chamber of Commerce after she showed its president a list of 51 companies that have signed onto the Corporate Renewable Energy Buyers Principles. Barnett says, "If we are able to focus on our strengths and things in common, this might put Kansas more broadly on the map as an example for other states. If you want a free market, this is it exactly."

During Obama's two terms as president, the price of solar panels dropped more than 80 percent. On economics alone, solar has begun to compete with coal and natural gas as a power source, and over the first half of 2015 nearly 40 percent of the nation's new electrical capacity came from solar. Kansas is very much in that potential mix. "We're among the top ten states for solar resource opportunities," says Barnett. "But this will mostly come from corporate purchases, and we've got to change the policies to make that happen." IKEA, the world's largest home furnishing company and among the ten largest corporate solar users, announced in 2015 that Kansas' biggest solar array (92,000 square feet) will crown a recently-opened store in Kansas City's suburban Merriam.

Kansas has barely tapped energy efficiency opportunities; there simply aren't any standards. But a Kansas Energy Investment Act allowing utilities to make money on efficiency investments is in the works. Westar, the state's largest investor in utilities, will have 40 percent wind in its portfolio by the end of 2017. It's simply cost-effective.

Kansas residents are certainly no strangers to the impacts of a rapidly-changing climate. Since 2010, weather events have cost at least a billion dollars in combined losses. Following unprecedentedly high temperatures, in March 2016 the largest wildfire in the state's history swept across more than 4,000 acres of Kansas and Oklahoma, stopped only by an unseasonable Easter snowfall. Governor Brownback had by then declared an emergency state of disaster for five counties.

Rabbi Rieber warns that this is only the beginning. "People know there are changes in the weather. The growing season starts earlier, the pollinators aren't necessarily back yet, the ground is less stable when there's more evaporation—which happens when it's warmer. From a climate resilience point of view, we grow less than 10 percent of our fruits and vegetables locally. It's all imported, mostly from California. So as they become less predictable as a food producer, that endangers our ability to get food. That's part of the consequences of the commodity farming situation that we've built up, one that is very water and land and carbon-intensive. Are we going to just keep going like this until things collapse? Or retool our mode of doing things?"

"Look," Rieber goes on, "western Kansas is going to run out of water within 50 years because they're sucking the Ogallala aquifer dry at astronomical rates. In fact, even faster because the corporate farmers know it's running out. Everybody out there is going to be dryland farming. The people who have the money are just saying, 'Well, my kids aren't gonna live here anyway.' So these towns are gonna dry up and blow away—to grow subsidized corn, which is a thirsty crop and is artificially price supported."

The rabbi sees the climate-change battle as an existential one for the Koch brothers and their political hirelings. "Climate change is the issue that triggers all the things right-wingers hate, because it needs government intervention, international agreements, and markets to be messed with. Those are all 'religious principles.' That's why they have to deny the problem, because the solutions are intolerable. They don't talk about climate change but 'managing for the extremes.'"

Barnett concurs that "Kansas is already in a very serious three-year drought. Across the state ranchers are having to sell off cattle because they couldn't provide food and water for them." But she is not as pessimistic as Rieber. Even the state's conservative Farm Bureau, she points out, has been vocal in supporting wind energy as an economic driver for agriculture. At the beginning of the drought, Barnett brought together a diverse steering committee to talk about what innovations in water and energy could look like. Governor Brownback even came to hand out nine Energy Progress awards, ranging from "doing dryland farming or drip irrigation or solar charging stations for their wind pumping, or farmers using wind to diversify income."

As he passed his 80th year on earth, Charles Koch certainly showed no sign of changing his tune. He still insisted to the *Washington Post* in June 2016 that "the scientific method" was on his side, criticizing the climate change movement for "trying to shut down and shout down and punish anybody who wants to enter into debate about it and for promoting renewable energy 'through corporate welfare.'"

Octogenarians like Koch might not dwell on the future of life on earth, but activist Dorothy Barnett does worry a lot these days about her daughter and grandson. They recently moved from their Kansas farm to Tulsa, and she is afraid that "they'll eventually have to move north because they're not going to have access to water and it's going to be too miserably hot. I think about farms that have been in families for generations which will not be able to see that kind of future for their grandkids. That is why I keep doing this work. It's a matter of conscience."

Chapter 9

A Children's Crusade?

Following the election of Donald Trump, the Obama administration released a new Climate Science Special Report produced by an inter-agency coordinating body made up of 44 scientists from government agencies, laboratories and universities. "Humanity is conducting an unprecedented experiment with the Earth's climate system," it declared. Even if greenhouse gas emissions are reduced significantly, temperatures in the US are expected to keep rising for the rest of this still-nascent century.

Since work began on this book two years ago, the climate change crisis has become an increasingly alarming specter for the American public. Even half of all conservatives now believe that climate change is real, up 19 percent over the past two years. Yet what many scientists are calling the "climate emergency" continues to escalate. After 14 straight months of record-breaking high temperatures, in July 2016 the UN declared that the year would be the hottest ever recorded. Meantime, the latest inventory of greenhouse gas emissions shows that the amounts of carbon dioxide and methane flooding the atmosphere is still accelerating.

Arctic sea ice is at an all-time low and may shortly not exist at all. Fish everywhere are finding it harder to breathe, as the warming atmosphere depletes oxygen in the ocean. Extreme drought plagues much of the world, including California and other swaths of the United States. Millions of trees have died off all across Europe, the American Southwest and California. Massive wildfires forced evacuation of over 80,000 residents of Fort McMurray in Alberta, Canada, the hometown of tar sands oil.

And the predictions for the future are grimmer by the day. According to the World Bank, water shortages by 2050 will be bringing unprecedented conflicts and human migrations across the Middle East, North Africa, Central and South Asia. According to another World Bank report, 1.3 billion people, and $158 trillion in assets, are at risk by 2050 from rising sea levels and flooding. (Overall sea levels are anticipated to soar by six feet by 2100). What are called "feedback loops"—one trend exacerbating another—will result in more ocean acidification, more fires bringing intensifying heat, more dead trees releasing more carbon than they store, and melting tundra sending tons more methane skyward.

A report out of Great Britain states flatly that people may soon be five times more likely to perish in an extinction event than an automobile crash. In short, many of us may be toast.

Will our children be the "left-behind generation?" In the spring of 2016, the *Future of Children* periodical published by Princeton University and the Brookings Institution devoted an entire issue to "Children and Climate Change." 85 percent of the world's young live in developing countries, where they are already suffering the impacts of the climate crisis. "The rise in temperatures associated with carbon emissions is already damaging children's health and wellbeing," say the authors. It's not only that temperatures above 90 degrees Fahrenheit have an immediate effect on children's health, but that the number of days featuring such extremes has been increasing for several decades and will continue. "Excess heat while fetuses are still in the womb is associated with physical defects, delayed brain development, and various nervous system problems that affect later development. In older children, excess heat can directly reduce learning."

Then there is the spread of infectious diseases, including most recently the Zika virus transmitted by mosquitoes. Climate researchers at Oxford University say that warming has "now led to the widest geographic distribution of *Aedes aegypti* [the Zika carriers] ever recorded." In three-quarters of American cities, warming temperatures are shown to be lengthening the mosquito season. Higher temperatures expand opportunities for breeding in pools of stagnant water, make the mosquitoes hungrier and promote a more rapid growth of viruses in the insect's gut. The *Aedes aegypti* species takes only small portions of blood from several victims, thus affecting many more individuals.

And while the virus brings only flu-like symptoms to adults, it is devastating to fetuses, apparently causing infants to be born with microcephaly: with smaller heads, and brains, than normal. The first more than forty locally caused cases of Zika in the US were reported in Miami in summer 2016, and the Centers for Disease Control issued an unprecedented warning to pregnant women to avoid a particular Miami neighborhood. Nor will the epidemic be confined to southern Florida. It's been predicted that the mosquitoes' range will soon extend to traditionally cooler states such as Missouri, Kentucky, North Carolina and to Washington, DC.

As the world plunges toward the climate abyss, at the same time, there is a growing sense that forward-looking governments, communities and organizations are finally taking strong action and accelerating the shift from conventional fossil fuels to renewable energy. Solar power has already become the cheapest way to generate electricity. As of May 2016, jobs in the American solar energy sector outnumbered the total of workers in coal mining and the oil-and-gas industry put together. A report by the International Renewable Energy Agency (IRENA) says that solar also employs more women—nearly one-quarter of the 209,000 solar-related jobs in the US (though still well below the 47 percent figure of women in the overall economy). Worldwide, green energy employment increased by five percent in 2015, to a total of 8.1 million jobs (including 2.8 million in solar)—the majority of those in China, where the manufacture of solar panels is a national priority. IRENA estimates that, by 2030, the number of renewable energy jobs around the world will triple to some 24 million.

California created more solar jobs and installed more megawatts of capacity in 2015 than any other state. In fact, according to the Department of Energy, "for both utility-scale solar PV and solar thermal, California has more capacity than the rest of the country combined, with 52 percent and 73 percent of the nation's total, respectively."

The port of Los Angeles—the nation's largest in terms of container volume and cargo value—is erecting the first marine terminal in the world that will generate all of its energy from renewables. The green terminal is projected to eliminate more than 3,200 tons a year of greenhouse gases, equivalent to removing over 14,000 cars a day from the choked freeways of Southern California.

Meanwhile, in Oregon, Governor Kate Brown signed legislation in March 2016 aimed at eradicating the use of coal—which currently accounts for 30 percent of the state's electrical generation—within two decades. Utilities will be mandated to provide half their power from renewable resources by 2040. When the existing hydropower is factored in, this is likely to mean between a 70 and 90 percent carbon-free electricity sector over the next 25 years.

The notion of "zero net energy buildings" is catching on. These are homes and office buildings that consume only as much renewable energy as they create on site over the course of a year. Denver has thousands of these residences in development, utilizing combinations of rooftop solar panels, smart thermostats, advanced water heaters and other energy innovations. Other such housing developments are underway in New England, New York, and the Carolinas, with the Habitat for Humanity organization emphasizing such practices in its low-income construction projects.

Technological advances will hopefully speed up the process. Floating arrays of solar—"floatovoltaics," as one company has trademarked—are coming online in many places around the world. More than 50,000 of these, for example, float in Japan's Yamakura Dam reservoir, enough to power almost 5,000 homes. In the long run, the floating panels are considered more efficient than land-based counterparts because water cools them; they also restrict algae blooms and prevent water from evaporating, ideal for drought conditions.

In 2016, Norway announced it will become the first nation to commit to zero deforestation and intends to end the sale of fossil fuel-based cars by 2025. A Global Covenant of Mayors for Climate and Energy—representing over 7,100 cities in 119 countries—came together in an alliance to move toward a low-carbon future. This, according to former New York Mayor Michael Bloomberg, marks "a giant step forward in the work of achieving the goals that nations agreed to" at the Paris summit in December 2015. The latest global gathering, at Marrakech late in 2016, found country after country agreeing to implement and strengthen the Paris accords.

While it is critical to continue rapidly shifting our energy needs away from fossil fuels, about half of the carbon emissions are emanating from the land—deforestation, loss of wetlands, and soil-depleting agricultural practices. So what's known as "carbon farming" has begun to receive increasing attention. Photosynthesis, as every elementary school student learns, is what made life on earth possible for us. All our forest and soil carbon comes from photosynthesis, dating back eons when our atmosphere consisted primarily of carbon dioxide and very little oxygen. The first photosynthetic bacteria supported organisms that allowed the growth of trees and plants. Today, the process that made our planet livable for the multifarious community of creatures that inhabit the earth is being reversed not only by the clear-cutting of forests and the extraction of fossil fuels, but by the destruction of the very soil from where our food is grown.

It's imperative that agriculture be redesigned around the original definition of organic farming—using good practices to build healthy soil that doesn't need chemical fertilizers or pesticides. This requires plant cover, minimal tillage, species diversity and incorporating domestic animals into the crop rotation—allowing roots to grow deeper while soil and organic matter increase. More fertile soils could produce major crops at far less cost. And these carbon farming practices could also sequester a significant portion of our excess carbon dioxide, turning CO_2 into organic soil matter.

As the climate movement continues to grow, the "next-gen" inheritors of family fortunes are also becoming more active. Resource Generation, a New York-based nonprofit, has been engaged since 1998 in transforming family philanthropy among wealthy young Americans between 18 and 35. The group's new executive director, Jessie Spector, has made climate change a top priority. "So often, people on the progressive Left talk about divesting from fossil fuels. Great, but what are you *reinvesting* in?" she asks. "Maybe not just some big solar company, but small grassroots communities—shifting municipal systems to more sustainable and local control, making sure environmental solutions are not replicating the same sort of power structures that current energy systems have."

A Wesleyan graduate, Spector recalls inheriting a trust fund of several hundred thousand dollars from her grandparents when she turned twenty-one and "feeling quite paralyzed about what to do with my life that would be meaningful and impactful. I felt a lot of guilt around privilege." She discovered Resource Generation on the Internet, began as an intern there in 2008, and within five years had become executive director.

"We aim to show the intersection of the disproportionate impact of climate change on poor communities of color and developing countries," Spector continues. "It's about integrating climate justice through strategic planning toward equitable distribution of land, wealth and power. Not just about money, but about the control of resources."

One of Resource Generation's 400 members is tech-money heir Farhad Ebrahimi, whose Chorus Foundation specializes in climate-driven philanthropy. What's been happening until now, Ebrahimi says, was all "done in a vacuum—changes in individual behavior, market-driven energy efficiency retrofits, and a Beltway-centric push for policy." As the transition from fossil fuels occurs, he believes, it creates an opportunity to shift economic and political power in the direction of workers and local populations, instead of simply building a new clean energy infrastructure based on the same corporate power model.

So the Chorus Foundation has committed $10 million to grassroots groups over a 10-year period, including one in Appalachia that helps fund the transition of local electric co-ops from coal—long predominant in this eastern Kentucky area—to clean energy. In Buffalo, New York, Ebrahimi's foundation is working to revive the economy with green jobs and neighborhood development that doesn't displace people. They're also putting money into the Alaska Center for the Environment and Native American Rights Fund, to manage resources in a collaborative effort with indigenous Alaskan communities.

The battle lines are drawn, and more and more parents as well as young people are weighing in—and finding that their voices *do* count with the policy-makers. One of the main battlegrounds these days is the public school system, where the Koch brothers and other carbon energy titans are trying to flood classrooms with their corporate propaganda. Despite these corporate PR campaigns, the Next Generation Science Standards—common learning goals designed to support teaching that climate change is a clear and present danger—have now been adopted by more than half the states. In May 2016, the Public School Board in Portland, Oregon passed a resolution to "abandon the use of any adopted text material that is found to express doubt about the severity of the climate crisis or its root in human activities." The Portland school board took action after two local teachers pointed out many outdated passages about climate change in science textbooks, filled with qualifiers like "might," "may," and "could." The teachers then brought together parents and students to discuss how to "deal with this civilization-changing crisis."

Within days, as expected, the energy industry-funded Heartland Institute proclaimed the Portland school board's action was in effect a book ban, and alleged that students were being "indoctrinated instead of taught how the scientific method works." But this once-effective argument no longer held sway—not even when corporate propagandists tried it in places like coal-dependent West Virginia. There, the state board of education voted to withdraw its previously altered version of the Next Generation Science Standards and to stick with the scientific consensus on human activity as the cause of climate change—ignoring the screams from climate-change deniers, who called the board's action "a form of ideological child abuse" along the lines of "non-science" once practiced by the USSR, the Nazis, and Communist China.

New educational groups are springing up to resist the industry assault on science. Our Kids' Climate is "a growing international coalition of parent and grandparent groups from around the world who have come together to demand bold action to protect the children we love from catastrophic climate change." The US-based coalition includes organizations in Norway, Switzerland, Sweden, Finland ("Parents Roar"), Canada and Australia. One of the member groups is Climate Parents, based in San Francisco and founded by longtime labor and environmental campaigner Lisa Hoyos, which works to motivate parents "to advance the urgent and bold climate solutions our kids deserve."

The climate change movement is also moving beyond education to nonviolent civil disobedience. A recent survey reveals that one out of every six Americans (some 40 million adults) is willing to engage in direct action of this sort against corporate or government activities that exacerbate the planetary crisis. The Standing Rock protests, where thousands of people traveled to join the Sioux in protesting the $3.8 billion Dakota Access oil pipeline being constructed alongside their tribal land, are a prime example of direct action bringing results. In this case, the US Army Corps of Engineers called a halt to construction pending an Environmental Impact Statement, although the incoming Trump administration quickly ignored that edict.

Already, the judicial system has begun siding with climate activists in certain cases. In 2013, after two men blocked a ship from the unloading of 40,000 tons of coal at a power plant in Brayton Point, Massachusetts, they were initially charged on several counts, including disturbing the peace and conspiracy. The defendants, Jay O'Hara and Ken Ward, argued that they had no choice given the imminent threat posed by climate change. On the day their trial was to begin, the district attorney of Bristol County reduced the charges to a small fine, explaining: "Climate change is one of the gravest crises our planet has ever faced. In my humble opinion the political leadership on this issue has been gravely lacking." The D.A. then met with the accused and vowed to take part in a planned People's Climate March.

Five activists who came to be called the Delta Five did go to trial after blockading a train in Everett, Washington, shipping fracked oil from the Bakken field. They claimed their act was a "reasonable act of conscience" and that "we need to be shutting down our fossil fuel infrastructure and keeping that oil in the ground." A jury was permitted to hear testimony on their necessity defense, and found them guilty only of a trespassing misdemeanor. Outside the courtroom afterward, three jurors hugged the activists and said they'd join them at a climate lobby day in the state legislature.

On International Youth Day, August 12, 2015, what's since been described by climate activists Bill McKibben and Naomi Klein as "the most important lawsuit on the planet right now" was filed in the US District Court of Eugene, Oregon. Twenty-one young plaintiffs, ranging in age from 8 to 19 and joined by venerable NASA scientist Dr. James Hansen as a "guardian for future generations," charged the federal government with violating their fundamental constitutional rights to life, liberty and property by enabling the fossil fuel industry to continue destroying the earth's atmosphere. Our Children's Trust, a newly formed nonprofit based in Eugene, provided the attorneys to argue that the Constitution and the Public Trust Doctrine had been violated.

Not surprisingly, three industry trade associations soon moved to intervene in the milestone case. The American Fuel and Petrochemical Manufacturers represented ExxonMobil, Koch Industries, BP, Shell, and a host of others. The American Petroleum Institute also jumped in, on behalf of 625 oil and natural gas companies, as did the National Association of Manufacturers. They termed the unprecedented lawsuit "a direct, substantial threat to [their] businesses." In January 2016, a judge granted these corporate energy interests defendant status.

No one was quite prepared for the hundreds of citizens who showed up for the preliminary hearing that March. So many people gathered on the courthouse steps that the oral arguments were transmitted by video feed into three more Eugene courtrooms and another in Portland. The concerned citizens watched for two hours as US Magistrate Judge Thomas Coffin questioned Sean Duffy, the Justice Department attorney, about whether present-day energy benefits were being exchanged for future harm: "Are you robbing Peter to pay Paul?" Judge Coffin asked. "Both (water and air) are vital to life, right?"

"Yes, your honor," the attorney responded. Well, could the government sell Exxon the Pacific Ocean, the judge persisted? Duffy argued that yes, it could. In the closing argument, Our Children's Trust executive director and lead counsel Julia Olson asserted the young plaintiffs have the "right to prove the government's role in harming them has been knowing and deliberate" by ignoring "undisputed scientific evidence" known for many years.

At a press conference that followed on the courthouse steps, 16-year-old Victoria Barrett told a large crowd, "The future of our generation is at stake. People label our generation as dreamers, but hope is not the only tool we have. . . . I want to do what I love and live a life full of opportunities. I want the generation that follows to have the same and I absolutely refuse to let our government's harmful action, corporate greed and the pure denial of climate science get in the way of that."

A month later, in mid-April, Judge Coffin allowed the children's crusade to move forward, ruling against the motion to dismiss by the federal government and the fossil fuel industry. The judge called this an "unprecedented lawsuit" that went to the heart of "government action and inaction" which resulted "in carbon pollution of the atmosphere, climate destabilization, and ocean acidification." While rejecting the argument that Congress could peddle the coastal sea waters to Exxon, the judge upheld the youth attorney's constitutional claims, agreeing that young people today had been denied "protection afforded to previous generations."

Judge Coffin wrote that in view of elected representatives' continued inaction, it was the responsibility of the judiciary to step in: "The intractability of the debates before Congress and state legislatures, and the alleged valuing of short term economic interest despite the cost to human life, necessitates a need for the courts to evaluate the constitutional parameters of the action or inaction taken by the government. This is especially true when such harms have an alleged disparate impact on a discrete class of society."

One of the young plaintiffs, Xiuhtezcatl Tonatiuh Martinez, from the Earth Guardians organization in Colorado, has emerged as a particularly eloquent spokesperson for the youth campaign. After the judge's decision, the teenager declared, "We will not be silent, we will not go unnoticed and we are ready to stand to protect everything our 'leaders' have failed to fight for. They are afraid of the power we have to create change. And this change we are creating will go down in history." A hip-hop artist raised in the Aztec tradition, Martinez has also addressed the United Nations' General Assembly on behalf of his generation.

Two days after the presidential election, federal district court Judge Ann Aiken ruled in the Oregon case. Upholding public trust and due process rights under the Constitution, she allowed the case to go forward over the oil companies' objections, saying: "I have no doubt that the right to a climate system capable of sustaining human life is fundamental to a free and ordered society." Public trust rights "both predated the Constitution and are secured by it," and can't be "legislated away."

Our Children's Trust hasn't stopped with the federal lawsuit, branching out to help kids file climate cases in all 50 states. In May 2016, four youth plaintiffs convinced the Massachusetts Supreme Judicial Court to demand the state Department of Environmental Protection to "address multiple sources or categories of greenhouse gas emissions" and "set emission limits for each year" in order to meet mandated federal EPA standards by 2020. Eight young petitioners in the Seattle area also received a favorable ruling that pushed Washington's Department of Ecology to release an emissions rule by the close of 2016.

Meantime, in late January, Our Children's Trust lawyers delivered an ultimatum to the federal government. Following media reports about a takedown of information on the EPA's climate change website, they stated that the Justice Department was "required by law to preserve all documents, including electronically stored information," that could prove relevant to the youth plaintiffs' case. "Destroying evidence is illegal."

A graduate student at Eastern Michigan University, John Rozsa, took matters into his own hands. Between classes and a full-time job, he started downloading the pre-Trump information about climate change on the EPA's website into a new one he called EPA Data Dump. With close to 30,000 files available by early March, Rozsa's website had seen so much traffic—more than 200,000 users—that its server almost crashed.

Meantime, attorneys for Our Children's Trust filed a request for documents from the American Petroleum Institute. They're out to learn what the industry knew, and when they knew it, from the early research into the changing climate—and how they decided to sow the seeds of doubt. The kids are looking to further unveil the cover-up.

Is a day of reckoning at hand? Despite the increasingly frantic efforts of the Tillersons and Kochs, and their political henchmen—men who would sacrifice civilization for the sake of lining their pockets—it looks like among a majority of the public, the horsemen of the apocalypse are finally losing the climate change debate. But they are clearly determined to fight to the last shale oil field and final coal mine. And they continue to spend millions upon millions of dollars to slow climate change action, until they can extract every possible barrel of oil from the earth.

These dark lords like to pose as good family men, benefactors of charities and the arts, upstanding pillars of their community. But first and foremost they are enemies of life on earth. This book has sought to put a face to the entrenched evil that has pushed us to the point of no return. As freak storms and vast ice melts and killer droughts and weird temperature spikes plague the earth with rising frequency, it's important to remind ourselves that these calamities are the result of human activity. And among those most responsible for this destructive activity are the men profiled in this book.

But, fortunately for the future of life on the planet, there is another type of human activity at work today. And it is growing stronger by the day—in the streets, in classrooms, in courtrooms, in government offices, and even in the luxurious homes where the families of energy moguls live, and where the recognition is growing among the children and grandchildren of these men that their family wealth cannot entirely protect them from the furies of nature. This type of human activity—in service of life—must continue to grow. It is our only hope.